Mind Without Soul, Morality Without God

How the Brain Creates the Self and Humanity Creates Ethics (Book 5)

Dr Prabhat Das

MD, PhD, FACP

No God Required

How Science Explains Reality Without the Divine

Hardcover ISBN: 979-8-951067-12-8

Paperback ISBN: 979-8-951067-11-1

eBook ISBN: 979-8-951067-10-4

Printed in the United States of America

TABLE OF CONTENTS:

Author's Note

Why I Wrote This Series

Some books begin in libraries. Some begin in classrooms. This series began much earlier, in the mind of a curious boy growing up in India.

I was born into a Hindu religious family. Religion was not a distant theory in my childhood. It was part of daily life. It was in prayers, rituals, festivals, stories, customs, family conversations, and the quiet assumptions that surrounded me. God was not discussed as a possibility. God was treated as reality.

My father was a firm believer in an omnipotent God. To him, God was real, powerful, protective, and deeply involved in human life. His belief was sincere. It was part of his strength, his discipline, and his view of the world. I grew up respecting that seriousness.

But from my teenage years, another force began to grow inside me. Curiosity.

I wanted to know why things happened. I wanted to understand nature, disease, death, the stars, the human body, and the mind. I was not satisfied with an answer simply because it was old, sacred, or repeated by many people. I wanted to know whether it was true.

In high school, I became deeply impressed by the lives and writings of great scientists. I still remember going to the local library and borrowing books written by, or about, Newton, Darwin, Einstein, and others. Those books were not popular among students at that time. In fact, I may have been the first student to borrow some of them. They were covered with dust. Some had even been damaged by insects.

But to me, they were treasures. Those dusty books opened a new world.

Newton showed me that nature follows laws.

Darwin showed me that life has a history far deeper than ancient stories.

Einstein showed me that reality is stranger, deeper, and more beautiful than ordinary imagination.

These thinkers did not ask me to believe blindly. They invited me to understand. They did not frighten me with punishment. They opened the door to wonder. They showed me that truth does not become weaker when questioned. It becomes stronger.

That invitation changed my life.

My curiosity did not end with school. It continued through medical college, scientific training, clinical practice, and decades of work as a physician. In medicine, I learned the value of evidence. A diagnosis cannot be accepted because it feels comforting. A treatment cannot be trusted because it is traditional. A belief about the body must face reality. It must face observation, testing, correction, and outcome.

That same habit of mind gradually shaped the way I looked at religion & God.

Even now, at retirement age, I continue to read almost every day. Science, philosophy, history, religion, psychology, evolution, cosmology, neuroscience, and the human mind still fascinate me. The more I read, the more I see how far human knowledge has come. Many things once explained by God are now understood through natural causes. Thunder, disease, eclipse, famine, fertility, mental illness, death, life, and the vast universe itself no longer need supernatural explanation.

This book series, "**No God Required: How Science Explains Reality Without the Divine", is a 7 book series**, and is my attempt to share that long journey in clear and simple language.

I do not write these books to insult believers. Many believers are kind, sincere, moral, generous, and deeply thoughtful people. I know this from my own family, my own culture, my patients, my friends, and my life experience. Faith often gives people comfort, hope, discipline, community, and emotional strength.

But faith does not decide truth.

A belief may comfort us and still need examination. A tradition may be ancient and still be mistaken. A prayer may bring peace and still not prove that anyone is listening. A sacred story may be beautiful and still be human-made.

My goal is to understand how belief began, why it became powerful, and why it still feels so real to so many people.

Why did human beings create gods?

Why did religion become one of the strongest forces in history?

Why do beliefs learned in childhood feel like truth?

Why do different cultures create different gods?

Why did science gradually make divine explanations unnecessary?

Can human beings live meaningful, moral, compassionate, and courageous lives without God?

These are serious questions. They deserve serious answers.

This series is based on knowledge gathered from many great books, scholars, scientists, historians, philosophers, psychologists, and thinkers. I have tried to bring their insights together in language that an ordinary educated reader can understand. My hope is that these books will encourage readers to think more rationally, question inherited beliefs, respect evidence, and study these subjects more deeply.

I also hope they will show that life without God does not have to be empty, cold, selfish, or immoral.

A life guided by reason, compassion, truth, science, justice, and human responsibility can be deeply meaningful. We do not need divine command to care for one another. We do not need fear of hell to be moral. We do not need supernatural stories to feel wonder before the universe.

Human beings are already capable of love, courage, kindness, beauty, and truth.

The universe is already astonishing.

Life is already precious.

Our responsibility to one another is already urgent.

No God is required for that.

Introduction: If the Soul Is Real, Why Does the Brain Control the Mind?

A stroke can steal speech. A tumor can bend judgment. Dementia can slowly erase familiar names, beloved faces, and the warm map of a lifetime. Alcohol can loosen restraint. A head injury can change impulse, anger, affection, and personality. Anesthetic drugs can switch off consciousness so completely that the world vanishes, and then, when the brain wakes again, the world returns.

These are not small medical curiosities. They are clues to one of the deepest questions human beings have ever asked.

What are we?

Are we immortal souls temporarily living inside bodies? Or are we living bodies whose brains create memory, personality, awareness, love, fear, conscience, imagination, and the feeling of being a self?

For thousands of years, religion answered with confidence. It said that a soul lives inside the body. The soul thinks. The soul remembers. The soul chooses. The soul loves. The soul sins. The soul survives death. The body may grow old and decay, but the soul remains.

That belief gave comfort to millions. It softened the terror of death. It gave grief a promise. It made human beings feel that something invisible and eternal lived inside them. It also gave morality a heavenly authority. If the soul would be judged after death, then every action carried eternal weight.

But comfort is not the same as truth.

A belief may be ancient and still be wrong. It may be emotionally powerful and still be unsupported. It may be repeated by billions and still fail when placed against evidence.

The serious question is not whether the idea of the soul feels comforting. The serious question is whether it explains what we actually see.

And what we actually see is clear: when the brain changes, the mind changes.

When parts of the brain are injured, speech may disappear. When brain circuits degenerate, memory may fade. When brain chemistry changes, mood and behavior may change. When anesthetic drugs alter brain activity, consciousness may vanish. When disease attacks the brain, personality, judgment, recognition, impulse, and awareness may all be affected.

If the soul is the true center of the person, why does the brain have so much power over the person?

This book begins with that question.

It does not begin with anger against religion. It does not begin with contempt for believers. It begins with evidence, human experience, and one honest challenge: if the soul explains the mind, then the soul should explain the facts better than the brain does.

But it does not.

The mind does not look like a ghost using the brain. It looks like what the living brain does.

The Last Shelters of God

In the earlier books of this series, we looked outward. We examined how human beings created gods from fear, death, nature, ignorance, and imagination. We looked at how science slowly replaced divine explanations. We looked at the universe without a creator and life without a designer.

Now we turn inward.

After gods retreat from thunder, disease, eclipses, stars, and the origin of life, many people still hold on to three final claims: science may explain the body, but it cannot explain the soul; science may explain nature, but it cannot explain morality; science may explain life, but it cannot give life meaning.

These claims are powerful because they do not touch only the intellect. They touch the heart.

The soul is not just a doctrine. For many people, it is tied to the hope that death is not final. It is tied to the longing that a dead parent, spouse, child, or friend may still exist somewhere. It is tied to the feeling that the person inside the body must be more than flesh, blood, nerves, and brain cells.

Morality is also not just a theory. It is tied to guilt, justice, punishment, love, children, family, kindness, cruelty, and the hope that good and evil are not merely personal opinions.

Meaning is not just a philosophical word. It is tied to the fear that if there is no heaven, then love, sacrifice, suffering, courage, and kindness may somehow become empty.

So religion says: without a soul, you are nothing; without God, morality collapses; without heaven, life has no meaning.

This book answers those fears.

It argues that we do not need a soul to explain the mind. We do not need God to explain morality. We do not need heaven to make life meaningful.

This is not a cold view of human life. It is not a reduction of human dignity. It is a more honest foundation for dignity. We are not less human because the brain creates the mind. We are more understandable. We are not less moral because ethics grows from human life. We are more responsible. We are not less meaningful because life is finite. We are more awake to the value of the time we have.

The Journey of This Book : This book moves through three connected questions.

The first question is about the mind: **Do we need a soul to explain the self?**

The opening chapters examine how human beings invented the soul from breath, dreams, death, grief, and fear. They then look at what modern knowledge of the brain reveals. Brain injury, dementia, drugs, sleep, anesthesia, coma, memory, emotion, personality, and consciousness all point toward the same conclusion: the mind is not a ghost inside the body. The mind is what the living brain does.

The second question is about morality: **Do we need God to be good?**

The middle chapters examine the roots of morality in care, cooperation, fairness, empathy, suffering, and human social life. They answer the common religious fear that without God, anything is permitted. They also examine the danger of divine-command morality, where cruelty can be protected if it is called sacred. A better morality does not begin with obedience. It begins with conscious beings who can suffer, flourish, be harmed, be protected, be humiliated, or be freed.

The third question is about meaning: **Do we need heaven for life to matter?**

The later chapters examine death, grief, love, legacy, responsibility, and the meaning of a finite life. Religion says that without immortality, life becomes empty. But a flower is not worthless because it fades. A song is not meaningless because it ends. A human life is not empty because it is brief. Meaning does not require eternity. It requires value.

These three questions form the heart of this book.

Mind without soul.

Morality without God.

Meaning without immortality.

Together, they lead to one central answer: the human being does not disappear when God disappears. The self remains real. Goodness remains real. Love remains real. Suffering remains real. Responsibility remains real. Meaning remains real.

What disappears is not humanity.

What disappears is the need for supernatural explanation.

The Brain Reveals What the Soul Cannot

The brain is not merely an organ inside the skull. It is the living center of mental life. It helps create perception, attention, memory, emotion, language, movement, impulse control, imagination, judgment, and awareness. It receives signals from the body and the outside world. It builds a model of reality. It stores fragments of the past. It prepares action. It gives us the feeling that "I am here" and "this is me."

This does not make human life less beautiful. It makes it more astonishing.

Think of a familiar face. Light enters the eyes. The brain organizes shapes, colors, lines, movement, and expression. But you do not experience a pile of visual details. You experience a person. In one instant, the face becomes your child, your friend, your spouse, your patient, your teacher, or your mother.

Think of music. Vibrations enter the ear. The brain turns them into rhythm, melody, sadness, joy, memory, and sometimes tears. Understanding sound waves does not destroy music. Understanding the brain does not destroy the beauty of music either.

Think of love. Love has biology, chemistry, memory, attachment, touch, time, culture, and choice inside it. But that does not make love fake. A rose is not fake because it has roots. A rainbow is not fake because light can be explained. A mother's love is not fake because the brain participates in it.

Science does not destroy wonder. It removes false explanations and leaves us with a deeper wonder.

The soul was supposed to explain the person. But the brain explains the person more directly, more consistently, and with far better evidence.

This becomes especially clear in illness. A person with damage to certain brain regions may lose the ability to speak, though intelligence and emotion may remain alive inside. Another person may recognize an object but not name it. Another may see but not understand what is seen. Another may lose emotional control. Another may become suspicious, impulsive, apathetic, or unusually childlike. Another may forget the spouse they loved for fifty years.

These changes are painful. They must be discussed with tenderness and respect. A person with dementia, stroke, psychiatric illness, addiction, or brain injury does not lose human worth. Human dignity does not depend on perfect memory, perfect speech, perfect behavior, or perfect reasoning.

But these conditions reveal something profound.

The patient does not lose dignity when the brain is damaged. The soul loses explanatory power.

If the soul were the true source of memory, judgment, personality, and awareness, then brain disease should not be able to alter them so deeply. Yet it does. That is one of the strongest reasons to doubt the soul.

The Fear That Goodness Dies With God

Even when some people begin to accept that the brain creates the mind, they may still resist atheism for another reason. They fear that without God, morality disappears.

This fear is repeated so often that many people accept it without serious examination. If there is no God, why be good? If there is no heaven, why sacrifice? If there is no hell, why not commit any crime? If morality is human-made, does that mean anything goes?

These questions sound serious, but they often hide a deeper problem.

If the only reason a person avoids cruelty is fear of divine punishment, that is not deep morality. That is fear. If the only reason a person tells the truth is hope of heavenly reward, that is not the highest form of honesty. That is a transaction. If the only reason a person does not harm a child, betray a friend, abuse the weak, or exploit the poor is because God is watching, then the moral problem is not atheism. The moral problem is a morality that never grew beyond surveillance.

Human goodness does not begin with theology.

A mother does not need scripture before she protects her child. A child does not need a priest before he cries over unfairness. A friend does not need a holy book before feeling the pain of betrayal. A stranger does not need a doctrine of heaven before pulling another person from danger.

Care, attachment, empathy, fairness, gratitude, guilt, shame, cooperation, and responsibility are woven into human social life. They are not perfect. They are not automatic. They can be distorted by fear, tribe, power, propaganda, prejudice, and religion itself. But they are real.

Morality grows because conscious beings can suffer. It grows because children need protection, trust matters, cooperation makes life possible, cruelty leaves wounds, and reason allows us to ask, "What if that were me?"

Religion did not create all this. Religion often claimed it. Religion sometimes shaped it. Religion sometimes protected it. Religion also often distorted it.

Morality is older and deeper than organized religion. This is why nonbelievers can be kind, generous, honest, brave, loving, and just. It is also why religious believers can be cruel, prejudiced, violent, or unjust when they surrender conscience to sacred authority.

The issue is not whether a person is religious or nonreligious. The issue is whether morality is built on fear, obedience, tribe, and ancient command, or on suffering, dignity, fairness, evidence, compassion, and human well-being.

When Obedience Pretends to Be Morality

Religion often says that morality comes from God, but this creates a serious problem.

Is something good because God commands it? Or does God command it because it is good?

If something is good only because God commands it, then morality becomes obedience to power. In that case, anything could become moral if God commanded it. Cruelty could become kindness. Slavery could become justice. Persecution could become duty. Killing could become holy.

But if God commands something because it is already good, then goodness does not depend on God. It exists independently, and human beings must still use moral judgment to recognize it.

This is not a small philosophical puzzle. It is one of the deepest problems in religious morality.

Sacred texts have been used to defend slavery, caste hierarchy, patriarchy, persecution, blasphemy laws, religious violence, and the silencing of doubt. Good people have often had to resist religious authority to expand human rights.

This does not mean every religious person is immoral. That would be false and unfair. Many religious people are kind, generous, gentle, honest, and devoted to service.

But when religious people are good, they are often good because of their humanity, not because of the harshest parts of their scriptures. They interpret, soften, ignore, or reject cruel passages. They choose compassion over literal obedience. They choose conscience over command.

In doing so, they quietly prove the point.

Moral judgment must stand above scripture. A command is not moral merely because it is ancient. A rule is not moral merely because it is sacred. A cruelty is not moral merely because it is religious.

A better morality asks better questions. Does this reduce suffering? Does this protect dignity? Does this respect freedom? Does this treat persons as persons? Does this stand up to evidence? Can it admit error? Can it improve?

That kind of morality does not need God. It needs honesty, compassion, courage, and human beings willing to think.

What Is Lost and What Is Gained

Some readers may feel that removing the soul takes something precious away. That feeling is understandable. The soul has been tied to childhood teachings, funeral prayers, family memory, hope for reunion, and the private feeling that something permanent lives inside us.

But what exactly is lost?

If there is no soul, love does not disappear. Grief does not disappear. Beauty does not disappear. Music does not disappear. Kindness does not disappear. The suffering of a child still matters. The dignity of a patient still matters. The rights of a stranger still matter. Truth still matters.

What is lost is not humanity. What is lost is an ancient explanation.

And what is gained?

We gain a clearer understanding of the brain. We gain compassion for people whose behavior is shaped by illness, trauma, addiction, dementia, depression, or brain injury. We gain a morality that does not depend on ancient commands. We gain responsibility for this world instead of waiting for another one. We gain urgency, because this life is not a rehearsal.

We gain honesty.

A false comfort may soothe for a while, but truth gives us firmer ground. There is dignity in facing reality without inventing another world to escape it. There is courage in accepting that the mind is fragile and life is brief. There is compassion in knowing that human beings are biological, wounded, social, dependent, and mortal. There is moral seriousness in understanding that no god will come down to stop cruelty, feed the hungry, protect the weak, cure disease, or defend justice.

If help is to come, it must come through us.

Responsibility, not despair, is the honest answer.

The Argument of This Book

Religion says the soul makes us human.

This book will show that the brain creates the self.

Religion says God makes us moral.

This book will show that humanity creates ethics.

Religion says heaven makes life meaningful.

This book will show that this one real life is meaningful enough.

The goal is not to make life smaller. The goal is to see life clearly.

We are not ghosts trapped in flesh. We are living beings made of matter, but matter organized in such an astonishing way that it can think about the stars, remember the dead, write poetry, comfort the sick, fight injustice, and ask why anything exists at all.

We are not puppets of heaven. We are social animals with the capacity for empathy, cruelty, cooperation, selfishness, courage, prejudice, reason, love, and moral growth.

We are not waiting-room creatures whose real life begins after death. We are mortal beings whose time is limited, whose choices matter, and whose kindness may be the only heaven another person ever knows.

That is not a poor view of humanity. It is a serious one.

It does not place human dignity in an invisible soul. It places human dignity in the reality of conscious life. It does not place morality in obedience to divine command. It places morality in the fact that suffering is real and compassion is possible. It does not place meaning in another world. It places meaning in this world, where children are born, people fall in love, patients suffer, friends forgive, strangers help, books are written, truth is discovered, and justice must be defended.

No soul is needed for any of this.

No God is required.

Transition to the Next Chapter

To understand why the soul became such a powerful idea, we must first return to the world before neuroscience.

Imagine early human beings watching a loved one die. The body is still there, but the person seems gone. The voice is silent. The eyes no longer answer. The breath has stopped. The warmth is fading. Then, days later, that loved one appears in a dream, speaking, smiling, alive again.

From such experiences, the idea of the soul was born.

The next chapter begins there: with breath, dreams, death, fear, and the ancient human attempt to explain the invisible self.

Sources

Antonio Damasio, *Descartes' Error: Emotion, Reason, and the Human Brain.*

Antonio Damasio, *Self Comes to Mind: Constructing the Conscious Brain.*

Anil Seth, *Being You: A New Science of Consciousness.*

Stanislas Dehaene, *Consciousness and the Brain.*

Michael Gazzaniga, *Who's in Charge? Free Will and the Science of the Brain.*

Michael Tomasello, *A Natural History of Human Morality.*

Frans de Waal, *The Bonobo and the Atheist.*

Internet Encyclopedia of Philosophy, "Divine Command Theory."

Stanford Encyclopedia of Philosophy, "Religion and Morality" and "Theological Voluntarism."

Chapter 1: The Ancient Invention of the Soul

How Breath, Dreams, Death, and Fear Created an Invisible Self

The fire is low. The night is cold. A small group sits close together, watching an old man die.

His breathing has become slow and uneven. His chest rises, pauses, then rises again. His family waits in silence. A child watches from the corner, frightened by what he does not understand.

The old man's face is still the same face. His hands are the same hands that once worked, held, fed, and blessed. His voice, only yesterday, still carried memory. But now the voice is gone. The eyes no longer answer. The body is present, yet the person seems to be leaving.

Then comes the final breath. The chest becomes still. The warmth begins to fade. The people around him feel that something invisible has departed.

That night, one of them dreams of him. In the dream, the old man is alive again. He speaks. He smiles. He seems to have gone somewhere, yet still exists.

To the ancient human mind, this was not merely a dream. It felt like evidence. The body had died, but the person had returned in sleep. The breath had left, but the face had appeared again in the night. Something must have escaped the body. Something must have survived.

Human beings did not yet know about neurons, oxygen, brain death, memory networks, REM sleep, or the electrical activity of the nervous system. So they gave the mystery a name.

Soul.

Spirit.

Breath.

Life force.

Invisible self.

The soul was not discovered in a laboratory. It was not found under a microscope. It was not measured, weighed, photographed, or tested. It was imagined by human beings trying to explain the most emotional experiences of life: breath, dreams, death, grief, memory, personality, fear, and hope.

This does not mean ancient people were foolish. They were observant, vulnerable, loving, and afraid. They mourned their dead. They watched the living body become a corpse. They tried to understand why breath meant life and breathlessness meant death. They tried to explain why the dead seemed to return in dreams.

The soul was an early human answer to real human questions.

But an answer can be understandable and still be wrong.

That is the central point of this chapter: the soul was humanity's ancient explanation for the self before humanity understood the brain.

When Breath Looked Like Life

There is something mysterious about breath. We cannot hold it in our hands. We cannot see it clearly. Yet without it, life disappears quickly. A newborn announces life with a first breath. A dying person leaves life with a final breath. Between those two moments, every hour depends on that quiet movement of air.

It is easy to understand why ancient people connected breath with life. They saw the chest rise and fall. They heard breathing during sleep. They felt warm air leave the mouth. They saw that a living body breathed, while a dead body did not. Before physiology, breath looked like the invisible thread holding life to the body.

This connection appears in many ancient cultures and languages. Words for spirit, soul, wind, and breath often overlap. The Hebrew word *ruach* is connected with breath, wind, or spirit. The Greek *pneuma* can mean breath, spirit, or wind. The Latin *spiritus* carries a similar connection.

These words reveal something important. The soul was not first imagined as a precise theological doctrine. It began as a human interpretation of life itself. Breath was invisible. Wind was invisible. Life seemed invisible. So the unseen movement of air became linked with the unseen force of life.

This made emotional sense. A person dies. The body remains. The face remains. The hands remain. The eyes remain. But the chest no longer rises. The voice is gone. The warmth fades. Something is missing.

To people without modern biology, it seemed natural to imagine that the missing thing was an invisible life force. Something had entered the body at birth and departed at death.

But we now know more. Breath is not a supernatural visitor. It is part of a biological system. Oxygen enters the lungs, moves into the blood, reaches cells, and supports the chemistry of life. The brain depends intensely on oxygen. When oxygen is cut off, brain function rapidly fails. Consciousness fades. Brain injury follows. Death can occur.

Ancient people saw breath leave and imagined the soul leaving. The mistake was natural, but it was still a mistake.

Breath is not evidence of an immortal soul. Breath is evidence that living bodies need oxygen.

The soul began, in part, because the invisible act of breathing looked like an invisible self.

Dreams and the Wandering Self

Dreams must have been among the most powerful sources of soul-belief.

Even today, dreams can feel intensely real. In a dream, a person can run, speak, fear, love, fight, fly, fall, and grieve. A person can meet the dead. A person can return to childhood. A person can stand in a house that no longer exists. A person can speak to a parent who died many years earlier and wake with tears on the pillow.

Now imagine the ancient world: no neuroscience, no sleep laboratory, no knowledge of REM sleep, no understanding of memory networks, emotional processing, or brain activity during sleep.

A person goes to sleep in one place, but in dreams seems to travel somewhere else. The body lies still, but the dream-self moves. The eyes are closed, but another world appears. A dead person is buried, yet appears alive in the night.

What would an ancient person conclude?

It would be natural to think that something inside the body could leave, wander, visit other places, and meet the dead. That "something" became the soul.

Modern neuroscience gives us a different explanation. Dreams are not journeys of an invisible spirit. They are experiences generated by the sleeping brain. They often combine memory, emotion, fear, desire, grief, and imagination into vivid inner worlds.

Dreams are not meaningless. They can carry emotional truth. They can reveal longing, anxiety, trauma, love, guilt, and unfinished grief. They can bring back voices we miss and places we thought we had lost.

But they do not require a soul traveling outside the body. The dreaming brain is enough.

This matters because dreams still keep soul-belief alive. A grieving wife dreams of her dead husband and feels that he visited her. A mother dreams of her dead child and feels that the child is still alive somewhere. A son dreams of his dead father giving advice and feels that the father's spirit came back.

These experiences are deeply moving. They should not be mocked. Grief deserves tenderness. Love can continue in memory long after death. Dreams can bring comfort to a wounded heart.

But comfort is not proof.

A dream of the dead shows the power of memory, love, and the brain. It does not prove that a soul has returned from another world. The dead live in dreams because they first lived in the brain of the living.

That is not a small thing. It is touching, human, and real.

But it is not supernatural.

Death and the Hope That Someone Remains

Death is the great wound from which many religious beliefs were born.

Every human society has had to face it. Every family has known it. Every loving person has feared it. Death is not only a biological event. It is an emotional earthquake.

A person who laughed yesterday is silent today. A person who answered when called no longer answers. A person who held your hand cannot hold it back. The body is present, yet the person seems absent.

This is one of the most painful human experiences. It is also one of the most powerful sources of belief in the soul.

When someone we love dies, the mind resists finality. The heart says, "This cannot be all." Memory continues to speak. Love continues to reach. The dead person remains present in photographs, rooms, clothing, habits, stories, and dreams.

So human beings imagined continuation. The person must still exist somewhere. The body is gone, but the soul remains. The grave receives the flesh, but heaven receives the person.

This belief softened grief. It gave mourners a way to endure what felt unbearable. It allowed people to believe that separation was temporary, that justice would be done, and that love could survive death.

That is why soul-belief became so powerful. It answered not only curiosity, but pain.

Human societies do not simply dispose of bodies. They mourn, bury, burn, preserve, decorate, chant, pray, remember, and tell stories. Death rituals show that human beings need meaning around loss. They need a way to hold the dead inside the living world of memory and love.

But that need does not prove the soul.

There is a difference between a psychological need and a factual truth. We may need comfort, but comfort does not create reality. We may long to meet the dead again, but longing does not prove an afterlife. We may feel that love is too great to end, but feeling does not defeat biology.

Death made the soul emotionally necessary long before evidence made it credible.

That is why belief in the soul persists. It is not merely an idea. It is tied to grief, fear, love, and hope. To question the soul can feel, to many people, like taking away the dead.

But that is not what science does.

Science does not erase the dead from memory. It does not erase love. It does not erase the influence of a life that has touched other lives. It does not erase grief, gratitude, legacy, or tenderness.

Science simply says that we have no good evidence that an invisible immortal self leaves the body and continues elsewhere.

That may feel painful. But truth is not always designed to comfort us. Sometimes truth asks us to become stronger, more honest, and more deeply compassionate toward the fragile life we actually have.

The Body That Stayed and the Person Who Vanished

There is one experience that almost forces the soul into the human imagination: the body remains after death, but the person does not.

This is emotionally overwhelming. The face may still be recognizable. The hands may look familiar. The hair, scars, and shape of the body remain. But the voice is gone. The eyes no longer respond. The personality is absent. The warmth is fading.

To ancient people, it seemed as if the real person had left.

Even today, people often speak this way. We say, "He is gone," even when the body is still in the room. We say, "She has left us," though the body is present. We say, "That is only the body," because the living presence we loved is no longer there.

This language is natural, but it can mislead us.

The person has not left the body like a bird leaving a cage. The person has ceased because the living processes that created that person have stopped.

A flame is not a little creature that flies away when a candle is blown out. The flame ends when the conditions that sustain it end.

Music is not a ghost inside the violin. It appears when strings, motion, air, and hearing come together. When the playing stops, the music does not travel to another world. It ends, though its memory may remain in those who heard it.

The self may be more like flame or music than like a hidden object inside the body. It is real. It matters. It can be beautiful. But it may not be a separate immortal substance.

This is difficult to accept because we feel like a permanent "I" inside the body. We do not feel like a process. We feel like someone.

But feeling permanent does not prove permanence. The sun feels as if it moves around Earth. It does not. The ground feels still beneath our feet. It is not. A dream feels real while we are dreaming. It is not. The self feels like a soul. That does not mean it is one.

The Invisible Answer to Visible Mystery

The soul explained many things before science could. It explained why a body moved, why a person breathed, why dreams felt real, why the dead seemed to return in sleep, why a living person had personality, why people felt guilt, and why the body died while love seemed to remain.

The soul was a single invisible answer placed over many visible mysteries.

That is how many religious ideas begin. When human beings do not understand something, they often imagine an unseen agent behind it. Thunder becomes the voice of a god. Disease becomes punishment. Madness becomes possession. Fertility becomes blessing. Drought becomes divine anger. Dreams become messages. Death becomes the departure of the soul.

This pattern is understandable. The human brain searches for causes. When something important happens, we want to know why. If the cause is not visible, we imagine one. If the event is frightening, we often imagine an intentional force behind it.

A branch breaks in the forest. The mind asks, "Who is there?" A storm destroys crops. The mind asks, "Who is angry?" A child becomes sick. The mind asks, "Who caused this?" A loved one dies. The mind asks, "Where did they go?"

This habit may have helped survival. It is safer to mistake the wind for a predator than to mistake a predator for the wind. But this same habit can also fill the world with invisible beings: spirits in trees, gods in storms, demons in illness, souls in bodies.

This is why belief in souls, spirits, ancestors, ghosts, gods, and invisible powers appears across many cultures. Human beings everywhere faced death, dreams, fear, illness, and mystery. Human minds everywhere searched for agency and meaning.

One early theory of religion, associated with E. B. Tylor, connected animism and soul-belief with attempts to explain dreams, visions, death, and life. Modern anthropology has moved far beyond many older nineteenth-century assumptions, but one basic insight remains useful: early spiritual ideas often grew from human attempts to explain experiences that seemed to reveal an invisible self or life force.

The details differed from culture to culture. Some believed in one soul. Some believed in many parts of the soul. Some believed ancestors remained near the living. Some believed the dead traveled to another realm. Some believed spirits occupied animals, rivers, mountains, trees, or sacred places.

But the emotional root was often similar: the world felt alive, death felt incomplete, dreams felt like travel, and the body seemed inhabited by something invisible.

Today, we can understand why this happened. We can respect the human need behind it. But respect is not the same as belief.

We can understand why people invented the soul without accepting that the soul exists.

The Soul Gave Power to Religion

Once people believed in the soul, religion gained enormous power.

If the soul survives death, then someone may claim to know its future. If heaven and hell exist, then priests, prophets, scriptures, and religious authorities can claim to know the rules. If the soul can be saved or damned, then fear becomes a tool.

This is why the soul became central to many religions. It gave death a story. It gave morality a judge. It gave suffering a possible explanation. It gave religious authority power over the invisible future.

A ruler can control the body with law, prison, and force. But religion can claim to control the fate of the soul. That is a much deeper power.

A person may tolerate poverty, injustice, caste, patriarchy, or suffering if told that obedience will save the soul. A person may fear doubt if told that doubt endangers the soul. A person may obey cruel commands if told that disobedience risks eternal punishment. A person may accept misery in this life if promised reward in another.

This does not mean that every belief in the soul was created for control. That would be too simple. Many people believed in the soul because they feared death and loved the dead. Many found comfort in imagining that life continues.

But over time, soul-belief became one of religion's strongest instruments. It made invisible claims more powerful than visible suffering. It made eternal reward more important than present justice. It made obedience feel safer than honest doubt. It made people afraid not only of death, but of what might happen after death.

This fear still shapes millions of lives.

A child is told that disbelief may lead to hell. A young person is told that natural desire may stain the soul. A grieving person is told that only the faithful will see loved ones again. A doubter is told that honest questioning may cost eternal life.

This is a heavy burden to place on the human mind.

And it all depends on one assumption: that there is an immortal soul whose fate religion can describe.

If the soul was an ancient explanation rather than a proven reality, then much of this fear begins to lose its foundation.

Why an Ancient Guess Is Not Modern Proof

Many old ideas once seemed obvious.

It seemed obvious that Earth was still and the sun moved across the sky. It seemed obvious that diseases came from curses, demons, bad air, or divine punishment. It seemed obvious that kings ruled by divine right. It seemed obvious that women were naturally inferior. It seemed obvious that some people were born into higher and lower ranks by sacred order.

Many "obvious" ideas were later shown to be wrong.

The soul may be one of them.

It felt obvious because people experienced themselves from the inside. They felt thoughts, memories, desires, and emotions. They did not feel neurons. They did not feel synapses. They did not feel brain networks. So they imagined an inner self separate from the body.

But science often reveals that reality is different from how it feels. We feel still, yet Earth spins and moves through space. We feel that color belongs directly to objects, yet color is created through light, eyes, and brain. We feel that the self is a fixed inner thing, yet memory, personality, mood, awareness, and identity can all change with the brain.

This is why personal feeling cannot be the final proof of the soul.

The question is not, "Do I feel like I have a soul?" Many people do. The question is, "What best explains the evidence?"

Does the soul explain dreams better than the sleeping brain? Does the soul explain personality better than brain development, injury, chemistry, and experience? Does the soul explain memory better than neuroscience? Does the soul explain dementia better than brain degeneration? Does the soul explain anesthesia better than altered brain activity? Does the soul explain consciousness better than the living brain?

Again and again, the answer is no.

The brain explains more. The soul explains less.

The soul was useful when knowledge was limited. It gave ancient humanity a way to speak about invisible life, personal identity, and death. But the history of knowledge has repeatedly shown that supernatural explanations shrink as natural explanations grow.

Lightning did not need Zeus. Disease did not need demons. Evolution did not need a designer. The mind does not need a soul.

The Humanity Behind the Error

It is important to say this clearly: calling the soul an ancient invention does not mean insulting the people who believed in it.

Human beings created the idea of the soul because they were human. They loved, dreamed, feared death, missed the dead, wondered why a body could be alive one hour and lifeless the next, and felt guilt, memory, longing, and inner life. They were trying to understand themselves.

In that sense, soul-belief deserves sympathy as a human creation. It was born from grief, imagination, fear, and love.

But we do not honor humanity by preserving every ancient mistake. We honor humanity by learning. We honor the dead by telling the truth. We honor life by understanding it more clearly. We honor the mind by studying the brain that makes the mind possible.

The soul was once a bridge between ignorance and comfort.

Science now offers a different bridge: from mystery to understanding.

This bridge may not promise eternal life. It may not promise reunion after death. It may not promise that a hidden self escapes the body. But it gives us something precious: honesty.

It teaches us that life is fragile. It teaches us that the brain is precious. It teaches us that memory, personality, awareness, and love depend on living conditions that can be damaged, protected, healed, or lost. It teaches us to care for the body, because the mind is not separate from it. It teaches us to treat people with brain disease, mental illness, addiction, dementia, and trauma with compassion rather than superstition or blame.

It teaches us that this life matters because this life is not guaranteed by another one.

The soul was invented to protect human meaning from death. But perhaps meaning does not need protection from truth.

Perhaps meaning becomes deeper when we know how fragile it is.

The First Step Away From the Soul

The first step away from belief in the soul is not hatred of religion. It is understanding.

Understand why people believed. Understand what they were trying to explain. Understand why breath looked like spirit. Understand why dreams looked like travel. Understand why death looked like departure. Understand why grief created hope.

Then ask a harder question.

Now that we know more about the brain, do we still need the soul?

That question will guide the next chapters.

The soul was humanity's first answer to the mystery of the self, but first answers are not always final answers. A child may first believe thunder is anger in the sky. Later, the child learns about clouds, electricity, pressure, and lightning. The wonder does not disappear. It changes.

Humanity once believed that the soul explained the mind.

Now we are learning that the brain creates the mind.

The wonder does not disappear.

It deepens.

Chapter Summary

The idea of the soul was born from ancient human attempts to explain breath, dreams, death, memory, personality, conscience, and grief. Before modern science, it was natural to imagine that an invisible life force lived inside the body and departed at death.

Dreams made the soul seem even more believable because the sleeping person appeared to travel, meet the dead, and experience vivid inner worlds while the body remained still. Death gave the soul its deepest emotional power. The body remained, but the person seemed gone, so people imagined that the real person had left for another world.

Religion later used soul-belief to give moral authority to heaven, hell, salvation, sin, and obedience. But an ancient explanation is not modern proof. Today, dreams, memory, consciousness, personality, and death are better understood through the brain and body than through the soul.

The soul was understandable as an early human idea. It is no longer necessary as an explanation.

Key Takeaways

The soul was not discovered through evidence. It was imagined to explain experiences that ancient people could not understand.

Breath, dreams, death, and grief played a major role in making the soul seem real.

The emotional power of soul-belief comes largely from grief and fear of death.

Modern science explains many experiences once attributed to the soul through biology, sleep, memory, and brain function.

Understanding why humans invented the soul helps us move beyond it without mocking those who believed.

Transition to the Next Chapter

The soul began as an ancient explanation for life, death, dreams, and the invisible self. But if the soul is real, it should explain the mind better than the brain does.

The next chapter looks at the strongest evidence against the soul: what happens when the brain is injured, diseased, drugged, or changed.

If changing the brain changes the person, then the mind may not be a soul inside the body.

The mind may be what the brain itself creates.

Sources

Edward B. Tylor, *Primitive Culture.*

Antonio Damasio, *Descartes' Error: Emotion, Reason, and the Human Brain.*

Anil Seth, *Being You: A New Science of Consciousness.*

Patrick McNamara, *The Neuroscience of Religious Experience.*

E. J. Wamsley, "How the Brain Constructs Dreams," *eLife*, 2020.

National Center for Biotechnology Information, *The Possible Functions of REM Sleep and Dreaming.*

D. M. Varisco, "The End of Life, The Ends of Life: An Anthropological View," *Journal of the Islamic Medical Association of North America*, 2012.

R. E. Watson-Jones et al., "Does the Body Survive Death? Cultural Variation in Beliefs About Life Everlasting," *Cognitive Science*, 2016.

Chapter 2: The Brain Creates the Mind

What Injury, Dementia, Drugs, and Anesthesia Reveal

If the soul is the driver and the brain is only the vehicle, then damage to the brain should damage the vehicle, not the driver.

A broken steering wheel should not change the driver's personality. A damaged tire should not erase the driver's memory. A cracked windshield should not change the driver's judgment, language, restraint, or love.

But real life shows something very different.

When the brain changes, the person can change.

A stroke can leave a person unable to speak, though thoughts and emotions may remain alive inside. A tumor can press on the frontal brain and change judgment, impulse, or behavior. Dementia can slowly dissolve memory, language, recognition, and personality. Alcohol can turn a careful person reckless. Drugs can alter mood, desire, fear, and awareness. Anesthesia can make consciousness disappear for a time and then return.

These are not small facts. They are not merely medical curiosities. They strike at the heart of soul-belief.

If the soul is the true source of thought, memory, personality, language, conscience, and awareness, why do these human qualities depend so deeply on the condition of the brain?

This chapter looks at that question. It will not treat brain disease coldly. Stroke, dementia, addiction, trauma, psychiatric illness, and brain injury affect real people and real families. They bring suffering, confusion, grief, exhaustion, and love. A person with brain disease does not lose human dignity. A person with dementia is not less worthy. A patient with stroke, addiction, depression, or head injury deserves care, patience, and respect.

This is not an argument against dignity.

It is an argument against the soul.

Compassion does not require us to ignore what these conditions reveal. They reveal something profound: the mind is not floating above the body. The mind is deeply dependent on the brain.

The soul was once offered as an explanation for the person. But the brain explains the person more directly, more consistently, and with far better evidence.

The Problem Brain Damage Creates for the Soul

Imagine a man who was once calm, responsible, and thoughtful. Then he suffers an injury to the front part of the brain. Afterward, his family says, "He survived, but he is not the same."

He becomes impulsive. He makes poor decisions. He loses social restraint. He says things he would never have said before. His emotions change. His judgment weakens. The body is still there. The face is familiar. The voice may be the same. But something in the person's way of being has shifted.

What has changed?

If the soul were the seat of personality and moral character, then an injury to brain tissue should not so deeply alter those things. The soul should remain intact. The real person should remain untouched.

But clinical experience shows otherwise. Brain injury can change attention, memory, planning, inhibition, emotion, and social behavior. The frontal lobes are deeply involved in judgment, planning, restraint, decision-making, and personality expression. Damage in these areas can change not only what a person can do, but how a person behaves.

This does not mean that the injured person becomes less human. It does not mean that love, care, or dignity should be withdrawn. It means that the expression of the person depends on the brain.

That is a serious problem for the idea of a separate soul.

The famous case of Phineas Gage is often mentioned in this context. In 1848, Gage survived a severe injury when an iron rod passed through his skull and damaged parts of his frontal brain. The popular version of his story has sometimes been exaggerated, and we should be careful not to turn one case into a myth. Still, the case remains historically important because it helped draw attention to the connection between frontal brain damage and changes in personality, judgment, and social behavior.

But the larger point does not depend on one man. Modern neurology, neurosurgery, psychiatry, emergency medicine, intensive care, and rehabilitation medicine show again and again that brain damage can alter mental life.

A person may become apathetic after brain injury. Another may become disinhibited. Another may become irritable. Another may lose the ability to organize daily life. Another may understand words but struggle to produce them. Another may speak fluently but fail to make sense. Another may lose emotional warmth. Another may become suspicious or confused. Another may lose the ability to recognize familiar faces.

If the soul is supposed to explain the person, why does the person change when the brain changes?

The simplest answer is also the most powerful: the brain is not merely a container for the mind. The brain is the organ that creates the mind.

Stroke, Tumor, Trauma, and the Changed Person

A stroke can happen in a moment. A blood vessel is blocked or bursts. A part of the brain loses oxygen and nutrients. Brain cells are injured or die. Suddenly, a life changes.

One person cannot move one side of the body. Another cannot speak. Another cannot understand language. Another cannot swallow safely. Another cannot see part of the visual world. Another becomes emotionally unstable. Another loses awareness of one side of the body. Another seems present, yet cannot find words.

To the family, it can feel as if part of the person has been stolen.

But what has been stolen? Not by a demon. Not by a curse. Not by divine punishment. Not by a damaged soul. A part of the brain has been injured.

This is especially striking in aphasia, a language disorder often caused by stroke. A person with expressive aphasia may know what he wants to say but cannot produce the words. He may look frustrated, point, gesture, and struggle. The thought is there, but speech will not obey. Another person may produce words that flow easily but do not make sense. Another may hear language but fail to understand it. Another may lose reading or writing ability.

The soul did not forget language. The brain system that supports language was damaged.

This distinction matters. In earlier centuries, such changes might have been explained as possession, punishment, madness, or spiritual attack. Today, we know better. We can see how specific brain areas and networks are tied to language, movement, perception, memory, and behavior.

A tumor can reveal the same truth more slowly. A tumor growing in or near the brain may affect behavior before anyone knows it is there. A person may become forgetful, irritable, careless, depressed, disinhibited, or confused. Family members may first think it is stress, aging, stubbornness, sin, or moral weakness. Later, imaging may show a mass pressing on or invading brain tissue.

Again, the lesson is clear: change the brain, and the mind may change.

Head trauma gives another example. A fall, crash, blast injury, sports injury, or assault can injure brain tissue. Some effects are obvious: loss of consciousness, weakness, speech problems, seizures, confusion. Other effects are quieter: poor concentration, mood swings, impulsivity, depression, fatigue, memory trouble, anger, or personality change.

The person may look physically normal and still feel mentally altered. Families often say, "He came home, but he is different."

That sentence is heartbreaking. It is also philosophically important.

If personality, restraint, speech, memory, and judgment were produced by an independent soul, brain injury should not be able to reshape them so deeply.

But it can.

This does not make the injured person less human. It makes the brain more central to what we call human.

Dementia and the Fragility of Memory

Dementia is one of the most painful arguments against the soul because it touches love so directly.

A husband who once remembered every family story may begin asking the same question again and again. A mother may forget where she placed her keys, then forget appointments, then forget names, then forget the familiar layout of her own home. A person who was once gentle may become suspicious, angry, or fearful. A person who was once careful may become unsafe with money, cooking, driving, or medication. A person may look at a child, spouse, or friend and no longer recognize them.

For families, dementia is not an abstract disease. It is a slow grief. The body remains, but the familiar person seems to fade piece by piece.

Sometimes, even in that fading, moments of beauty remain. A woman who no longer remembers her daughter's name may still smile when an old song is played. A man who cannot follow conversation may become calm when his wife holds his hand. A patient who cannot recall the date may still respond to kindness in the voice of a caregiver.

These moments matter. They remind us that dignity does not depend on perfect memory. A person with dementia remains a person. Love remains. Respect remains. Care remains.

But as an explanation of mind, dementia creates a serious problem for soul-belief.

If the soul stores the real self, why does damage to the brain erase access to memory? If the soul is the center of personality, why can brain degeneration change behavior? If the soul is the true person, why can disease in brain tissue alter recognition, language, judgment, fear, mood, and identity?

Alzheimer's disease and other dementias affect brain cells, networks, and communication. They can damage memory, thinking, language, behavior, and daily functioning. Personality changes are common in many forms of dementia. Families may see apathy, irritability, anxiety, suspicion, emotional blunting, disinhibition, or loss of interest.

These changes do not mean the person has lost moral worth. They mean that the mind is vulnerable because the brain is vulnerable.

This truth should make us more compassionate, not less. A person with dementia is not choosing forgetfulness. A person with frontal lobe disease is not simply being rude. A person with brain degeneration is not morally failing by losing patience, judgment, or recognition.

Understanding the brain helps us move away from blame and toward care. Religion has often treated mental and behavioral changes as moral or spiritual problems. Science teaches a more humane lesson: the person is suffering because the brain is suffering.

This is one of the great moral gifts of neuroscience. It replaces superstition with understanding. It replaces condemnation with compassion. It replaces the language of sin, possession, and weakness with the language of illness, injury, vulnerability, and care.

Yet at the same time, dementia points to a hard truth. The self is not an untouchable soul hidden safely inside the body. The self is deeply dependent on memory, language, emotion, recognition, and brain function.

When these are damaged, the self can be altered.

That does not make the self unreal. It makes the self biological.

Drugs, Alcohol, and the Chemistry of Behavior

Alcohol is one of the most ordinary and familiar arguments against a separate soul.

A few drinks can change speech, judgment, coordination, impulse, emotion, and restraint. A quiet person may become talkative. A careful person may become careless. A gentle person may become angry. A shy person may become bold. A responsible person may take risks that he would normally avoid.

What happened to the soul? Did alcohol enter the soul? Did wine, whiskey, or beer chemically alter an immortal spirit?

Of course not. Alcohol changed the brain. It affected neural signaling, inhibition, judgment, coordination, and emotional regulation. The result was a changed pattern of behavior.

This is so common that we barely notice how philosophically important it is. If a chemical can alter mood, restraint, courage, aggression, memory, and decision-making, then mental life is not independent of biology.

The same is true for many drugs and medications. Sedatives can calm anxiety. Stimulants can increase alertness. Antidepressants can improve mood in many people. Antipsychotic medications can reduce hallucinations or delusions in some patients. Pain medications can change suffering. Psychedelic substances can produce intense changes in perception, emotion, and sense of self. Hormones can influence mood, energy, desire, irritability, and emotional sensitivity.

Even ordinary states such as hunger, sleep deprivation, fever, pain, and fatigue can change personality expression. A kind person may become irritable when exhausted. A calm person may become anxious after days without sleep. A patient in severe pain may appear angry, withdrawn, or hopeless.

The soul is supposed to be separate from the body. But daily life shows that the person we express depends heavily on the body.

This does not mean that human beings are helpless puppets of chemistry. That would be too simple. We can reflect, learn, build habits, seek treatment, make choices, and take responsibility. But our choices and character operate through the brain, not outside it.

The brain is not an enemy of the self. The brain is the condition that makes the self possible.

This is why medicine can change mental life. This is why sleep matters. This is why nutrition, pain, hormones, trauma, addiction, and medication matter. This is why psychiatric illness is not a defect of the soul. This is why brain health is human health.

A soul floating above the body cannot explain why a small molecule can change fear, joy, alertness, hallucination, memory, or restraint.

The brain can.

Anesthesia and the Switch of Awareness

General anesthesia is one of the most striking facts in all of medicine.

A patient lies on the operating table. The lights are bright. The monitors beep. The staff moves quietly. The anesthesiologist gives medication. Within moments, awareness fades. Pain disappears. Memory stops forming. The world is gone.

Then, after surgery, the drugs wear off. The eyes open. The world returns. The patient asks, "Is it over?"

To the patient, hours may feel like nothing. One moment there was the operating room. The next moment there is recovery.

Where was the soul during that time? If consciousness comes from an immortal soul, why can anesthetic drugs switch off experience?

Of course, anesthesia is medically complex. Not every anesthetic state is the same. Unresponsiveness is not always identical to complete absence of consciousness. Some brain activity continues. Different drugs act through different mechanisms. Modern anesthesia science is still studying exactly how anesthetic agents alter brain networks, communication, arousal, memory, and awareness.

But the basic fact remains powerful: changing brain activity can profoundly alter consciousness.

Anesthesia does not remove a soul from the body. It changes the conditions under which the brain produces awareness, memory, and response.

This is not speculation. It is daily medical reality. Millions of patients undergo surgery under anesthesia. Their consciousness can be reduced, interrupted, or profoundly altered by chemical effects on the nervous system. Then consciousness can return as the brain recovers from those effects.

That should make us pause.

Religion often speaks of consciousness as if it belongs to a soul. But anesthesia behaves exactly as we would expect if consciousness depends on brain function.

The same is true in the ICU. A patient may become more awake as sedation is reduced. Another may become confused when infection, low oxygen, kidney failure, liver failure, or medication affects the brain. Another may drift in and out of awareness depending on blood pressure, oxygen, fever, or metabolic state.

In these situations, awareness is not treated as a mysterious spiritual flame. It is treated as a brain function affected by the body's condition. That is why doctors check oxygen levels, blood sugar, electrolytes, infection, medications, and brain imaging. They are not looking for the soul. They are looking for the conditions that allow the brain to support consciousness.

A soul theory does not explain why awareness should depend so precisely on brain injury, drug state, oxygen, blood flow, metabolism, and neural networks. A brain-based theory does.

This does not make consciousness ordinary. It remains one of the great mysteries of science. But mystery is not magic. The fact that we do not yet understand every detail of consciousness does not mean that consciousness belongs to an immortal soul.

It means that the brain is more astonishing than ancient religion imagined.

The Clinical Evidence Against a Separate Soul

Consider the evidence together. Brain injury can change personality. Stroke can damage language. Dementia can weaken memory and recognition. Alcohol can alter judgment. Drugs can change mood, perception, and behavior. Anesthesia can interrupt consciousness. Sleep can change awareness. Pain, fever, hormones, trauma, and exhaustion can affect the way a person thinks and acts.

All these facts point in the same direction: the mind depends on the brain.

This is not an isolated argument. It is not based on one unusual case. It is the daily experience of medicine, neurology, psychiatry, surgery, emergency care, intensive care, rehabilitation, and family life.

Anyone who has watched a confused patient become clear after oxygen improves understands that awareness depends on brain function. Anyone who has seen a family struggle with dementia understands that memory is not stored in an untouchable soul. Anyone who has watched a patient wake from anesthesia has seen consciousness return with the recovery of brain function.

Every hospital is a quiet argument against the soul.

In the emergency room, a confused patient may become clear after oxygen is restored, blood sugar is corrected, infection is treated, or medication is adjusted. In the ICU, consciousness may rise and fall with sedation, oxygen, blood pressure, kidney failure, liver failure, infection, or brain swelling. In neurology, a small stroke may damage speech, vision, balance, or memory. In psychiatry, changes in brain function can produce depression, mania, hallucinations, anxiety, obsession, or disordered thinking. In geriatrics, dementia can slowly change the person a family has known for decades. In addiction medicine, substances can hijack reward pathways and weaken control.

In all these places, one lesson repeats: mental life is embodied.

The mind is not separate from the body. The person is not a ghost trapped inside flesh. The person is a living brain in a living body, shaped by genes, development, experience, relationships, culture, illness, injury, and time.

This view is not degrading. It is humane. It helps us understand why people change. It helps us treat illness. It helps us reduce blame. It helps us protect the brain. It helps us see the fragile beauty of the self.

The soul explains none of this with equal clarity.

Why the Brain Explains What the Soul Cannot

A good explanation should help us understand what happens. It should make sense of the facts. It should connect with other knowledge. It should be open to correction.

The brain-based explanation does this.

If a stroke damages a language area, speech may be affected. If dementia damages memory networks, memory may decline. If frontal systems are injured, impulse and judgment may change. If alcohol alters brain chemistry, behavior may change. If anesthetic drugs disrupt brain networks, consciousness may fade. If sleep overtakes the brain, awareness changes. If oxygen fails, consciousness can disappear.

These patterns are not random. They are connected to brain function.

The soul explanation does not work as well.

If the soul is separate from the brain, why should speech depend on brain tissue? If memory belongs to the soul, why should dementia erase it? If personality belongs to the soul, why should frontal injury change it? If consciousness belongs to the soul, why should anesthesia interrupt it? If moral judgment belongs to the soul, why should tumors, trauma, drugs, or degeneration affect it?

The usual religious answer is that the soul uses the brain like a musician uses an instrument. If the instrument is damaged, the music changes, but the musician remains untouched.

This analogy sounds attractive, but it fails.

A violin does not create the musician's memory. A piano does not create the pianist's personality. A broken flute may distort sound, but it does not make the musician forget his own child, lose moral restraint, develop hallucinations, or become unable to recognize his spouse.

Brain damage does not merely distort the expression of an unchanged soul. It can change the contents of mental life itself. It can change memory, desire, emotion, recognition, impulse, and awareness.

That is much more than a broken instrument.

A better analogy is fire. Fire is real, but it is not a separate object hidden inside wood. It is a process that appears when fuel, oxygen, heat, and chemical reaction come together. Change the conditions, and the flame changes. Remove the conditions, and the flame ends.

The mind may be similar. It is real, powerful, beautiful, frightening, loving, creative, and morally serious. But it may be a process of the living brain, not an immortal substance inside the body.

When the brain changes, the process changes. When the brain stops permanently, the process ends.

This may not be emotionally easy to accept, but it fits the evidence better than the soul.

A More Compassionate View of Human Beings

Some people fear that a brain-based view makes humans less worthy. They think that if we are biological, we are "only" machines.

That fear is understandable, but mistaken.

A brain-based view does not make a person less valuable. It makes the person more understandable. It tells us that the depressed person is not simply weak. The addicted person is not simply sinful. The patient with dementia is not simply difficult. The person with frontal brain injury is not simply immoral. The child with developmental problems is not simply disobedient. The traumatized adult is not simply broken by choice. The confused elderly patient is not possessed, cursed, or spiritually failed.

These are human beings whose brains, bodies, histories, and environments matter.

This view should deepen compassion. It should make us slower to condemn and quicker to understand. It should make us ask better questions.

What happened to this person? What is happening in the brain? What pain, illness, trauma, habit, addiction, stress, deprivation, or disease is shaping this behavior? What treatment, support, structure, protection, or rehabilitation may help?

This does not remove responsibility. People still make choices. Society still needs rules, accountability, and justice. Harm must be prevented. Victims must be protected. Dangerous behavior must be addressed.

But responsibility without understanding becomes cruelty.

A brain-based view allows justice to become more intelligent and more humane.

Religion often explains behavior through sin, temptation, possession, weakness, or disobedience. Sometimes those words may feel morally satisfying, but they often block understanding.

Science asks us to look deeper. Not to excuse everything. Not to deny harm. Not to remove accountability. But to understand causes.

Understanding causes is one of the first steps toward reducing suffering.

That is a moral advance.

What Remains Without the Soul

If the brain creates the mind, what remains?

Everything that matters remains: love, memory while the brain can sustain it, kindness, suffering, beauty, music, language, justice, and the need for compassion.

The fact that these arise from the brain does not make them false. A rainbow is not false because it depends on light and water droplets. A flower is not false because it depends on soil, sunlight, and genes. A song is not false because it depends on vibrating air and a hearing brain.

To explain something is not to destroy it. To understand the brain is not to cheapen the mind.

The soul was meant to protect human value by placing it outside nature. But perhaps human value does not need to be outside nature. Perhaps human value is inside nature.

Perhaps dignity belongs to conscious beings because they can suffer, love, hope, fear, remember, and care. Perhaps the person is precious not because an immortal soul hides inside the body, but because a living brain creates a fragile world of experience that can be harmed, healed, cherished, and lost.

This view does not make life meaningless. It makes life urgent.

If the mind depends on the brain, then the brain matters. Health matters. Safety matters. Childhood matters. Education matters. Nutrition matters. Sleep matters.

Addiction treatment matters. Mental health care matters. Dementia care matters. Stroke prevention matters. Trauma prevention matters. Compassion matters.

This is a serious view of humanity. It does not ask us to wait for heaven to value the person. It asks us to value the person here and now.

The Soul Shrinks as the Brain Is Understood

The soul once explained almost everything inside us. It explained thought, memory, personality, dreams, moral choice, consciousness, life, and death.

But with every advance in knowledge, the territory of the soul has become smaller.

Memory is studied through brain systems. Language is studied through brain networks. Personality is studied through development, temperament, biology, and experience. Emotion is studied through circuits, hormones, body states, and meaning. Dreams are studied through sleep science. Consciousness is studied through brain activity, arousal, attention, integration, and network function. Moral behavior is studied through empathy, evolution, social life, development, learning, culture, and reason.

The soul remains mostly where knowledge is incomplete. It survives in the gaps.

But the history of science teaches caution about gaps. Many things once placed in the gaps were later explained naturally. Thunder was not divine anger. Epilepsy was not demonic possession. Infection was not punishment. Evolution was not design. Mental illness was not a stained soul.

Again and again, God and soul were used as explanations before better explanations arrived. Then better explanations arrived.

The same pattern is happening with the mind.

We do not yet know everything about the brain. We do not have a complete explanation of consciousness. We do not fully understand every detail of memory, personality, or subjective experience. But incomplete knowledge is not evidence for an immortal soul.

A mystery is not a miracle.

A gap is not a ghost.

The honest path is to keep studying, not to fill ignorance with ancient answers.

Chapter Summary

This chapter examined the strongest evidence against the idea of a separate soul: the dependence of the mind on the brain.

Brain injury can change personality, judgment, impulse, emotion, and behavior. Stroke can affect speech, understanding, vision, movement, and awareness. Dementia can slowly damage memory, recognition, language, personality, and daily functioning. Alcohol and drugs can alter mood, restraint, perception, and judgment. Anesthesia can interrupt consciousness by changing brain function.

These facts are difficult to explain if the soul is the true source of thought, memory, personality, and awareness. They are much easier to explain if the mind is what the living brain does.

A brain-based view does not reduce human dignity. It can increase compassion by helping us understand illness, injury, addiction, trauma, depression, dementia, and behavioral change without superstition or blame.

The soul once explained the person when the brain was poorly understood. Modern evidence shows that the brain explains the mind far better than the soul does.

Key Takeaways

When the brain changes, the mind can change.

Brain injury, stroke, dementia, drugs, alcohol, and anesthesia all show the deep dependence of mental life on brain function.

The patient does not lose dignity when the brain is damaged. The soul loses explanatory power.

The soul does not explain why memory, language, personality, judgment, and consciousness can be altered by brain disease or injury.

A brain-based view of human beings supports compassion, treatment, prevention, and understanding.

The mind is real, but it does not need to be supernatural.

Transition to the Next Chapter

If the brain creates the mind, then the next question becomes even more personal.

What is the self?

We feel like one continuous person. We say "I" as if there is a fixed inner being who remains unchanged from childhood to old age. Yet memory changes, personality changes, the body changes, beliefs change, emotions change, and relationships change.

The next chapter will examine this mystery.

The self may not be an immortal soul hidden inside the body.

It may be a living story the brain keeps revising.

Sources

Antonio Damasio, *Descartes' Error: Emotion, Reason, and the Human Brain.*

Oliver Sacks, *The Man Who Mistook His Wife for a Hat.*

Anil Seth, *Being You: A New Science of Consciousness.*

Michael Gazzaniga, *Who's in Charge? Free Will and the Science of the Brain.*

Hanna Damasio et al., "The Return of Phineas Gage: Clues About the Brain from the Skull of a Famous Patient," *Science*, 1994.

P. H. M. de Freitas et al., "E.L., a Modern-Day Phineas Gage: Revisiting Frontal Lobe Injury," *Dementia & Neuropsychologia*, 2022.

S. M. Sheppard and S. Sebastian, "Diagnosing and Managing Post-Stroke Aphasia," *Expert Review of Neurotherapeutics*, 2020.

A. Terracciano et al., "Personality and Alzheimer's Disease: An Integrative Review," *Personality Disorders*, 2019.

J. Montupil et al., "The Nature of Consciousness in Anaesthesia," *British Journal of Anaesthesia*, 2023.

A. G. Hudetz, "General Anesthesia and Human Brain Connectivity," *Brain Connectivity*, 2012.

G. A. Mashour and M. T. Alkire, "Evolution of Consciousness: Phylogeny, Ontogeny, and Emergence from General Anesthesia," *Proceedings of the National Academy of Sciences*, 2013.

Chapter 3: The Self Is a Living Story

How Memory, Body, Emotion, and Relationships Create "Me"

You look at an old photograph and say, "That is me."

The child in the picture may have a different face, a different body, a different voice, different fears, different dreams, different beliefs, and a very different understanding of the world. Still, you say, "That was me."

But what exactly remained the same?

Your body changed. Your memories changed. Your beliefs changed. Your knowledge changed. Your emotions changed. Your relationships changed. Your hopes changed. Your fears changed. Even your brain changed.

Yet through all these changes, there remains a feeling of continuity. A thread seems to run from childhood to youth, from youth to adulthood, from adulthood to old age. You may no longer be the child in the photograph, but you do not feel like a completely different person either.

This feeling is powerful. It is one reason people believe in the soul.

The self feels like a permanent inner person, a private "I" hidden behind the eyes, watching life unfold through the body. The body grows older, but the "I" seems to remain. The world changes, but the "I" seems to continue. Thoughts come and go. Emotions rise and fall. Memories fade and return. But the "I" feels like the owner of all of them.

Religion takes this feeling and gives it a supernatural explanation. It says the self feels continuous because there is an immortal soul inside the body.

But there is another explanation.

The self may not be a fixed object. It may be a living story, one the brain keeps revising from memory, body, emotion, language, relationships, culture, and time.

A story can be real without being immortal. A life story can have continuity even as its details change. A childhood event may be understood one way at age twenty and another way at age sixty. A wound may first be remembered as shame and later understood as survival. A lost faith may first feel like emptiness and later become liberation.

The self is like that: real, but changing; continuous, but not fixed; powerful, but not supernatural.

The Feeling of a Permanent Inner Person

Every person lives from the inside.

You do not merely see your hand. You feel that it is your hand. You do not merely hear your voice. You feel that it is your voice. You do not merely have memories. You feel that they are your memories. You do not merely experience pain, hunger, fear, love, or joy. You feel that these things are happening to you.

This inner ownership is so familiar that we hardly notice it. It is the background of ordinary life. It creates the feeling that there is someone inside the body who receives experience, makes choices, and remains the same through time.

That feeling is real as an experience, but the explanation of that feeling may be wrong.

Human beings once felt that Earth was still and the sun moved across the sky. That feeling was powerful and natural, but it did not describe reality correctly. Human beings feel that color belongs directly to objects. A rose looks red. The sky looks blue. But color is not simply sitting inside objects. It is created through light, the eye, and the brain.

In the same way, the self feels like a fixed inner thing. But the feeling of a fixed self may be created by the brain.

This does not mean the self is fake. It means the self is constructed.

The brain gathers signals from the body, memories from the past, expectations about the future, emotions from the present, words from language, and responses from other people. From all this, it creates the experience of "me."

This is not a small achievement. It is one of the most astonishing achievements of the brain. But it does not require a soul.

A mirror can create a reflection. The reflection is real as an image, but it does not prove that another person is living inside the glass. In the same way, the brain creates a self. The self is real as experience, but it does not prove that an immortal soul is living inside the body.

Memory as the Thread of Identity

Memory is one of the strongest threads holding the self together.

Without memory, life would break into disconnected moments. You would wake each morning without a history. You would meet loved ones as strangers. You would not know your promises, your losses, your mistakes, your skills, your childhood, your work, or your place in the world.

Memory tells you where you have been. It allows you to say, "I was there. I did that. I loved that person. I suffered that loss. I made that choice. I learned that lesson."

This is why memory feels so close to identity.

But memory is not a perfect recording. It is not a video stored inside the brain. Memory is selective, emotional, and often reconstructed. It is influenced by later knowledge, repetition, desire, shame, pride, grief, fear, and other people's stories.

Two siblings may remember the same childhood event differently. A person may remember an argument in a way that protects self-respect. An old humiliation may grow larger with time. A joyful memory may become warmer each time it is retold. A painful memory may become sharper, or sometimes softer, depending on how life continues.

We do not merely retrieve memories. We rebuild them.

This matters because if memory helps create identity, and memory is changeable, then the self is changeable too.

The self is not a stone tablet. It is more like a manuscript revised over time. Some chapters are clear. Some are blurred. Some are rewritten. Some are missing. Some are interpreted differently after new experiences.

This does not make identity meaningless. It makes identity living.

A person who once saw himself as a failure may later see the same past as training, struggle, and growth. A person who once felt shame may later understand trauma. A person who once believed deeply in religion may later reinterpret life through science and reason. The facts of the past may remain, but the meaning of the past changes.

So the self changes.

If the soul were a fixed, unchanging essence, this would be difficult to explain. But if the self is a brain-based and memory-based process, this is exactly what we should expect.

We are not born with a completed self.

We become.

When Memory Breaks, the Self Is Shaken

The importance of memory becomes painfully clear when memory is damaged.

In dementia, the story of the self can become fragmented. A woman may remember songs from childhood but not the meal she just ate. A man may remember his village from sixty years ago but not recognize the room he is sitting in. A mother may forget that her children are adults. A husband may ask for his wife while she is standing beside him. A person may remember fragments of life but lose the thread that connects them.

Families often say, "She is still there, but she is fading."

This sentence carries both love and sorrow. It protects dignity, but it also expresses the truth that memory and identity are deeply connected.

If the self were an untouched soul, dementia should not be able to disturb identity so profoundly. Yet it can.

Again, this does not mean that the person with dementia loses human worth. Human dignity is not earned by memory, intelligence, independence, or perfect personality. A person with dementia remains a person. The duty of care remains. Love remains. Respect remains.

But dementia does reveal something about the self. The self depends on living brain systems. When memory networks are damaged, the story of the self can become fragmented.

This is not what we would expect if the real self were an immortal soul stored safely beyond the brain. It is exactly what we would expect if the self is created through the brain's fragile work of memory, emotion, recognition, and meaning.

The lesson is painful, but it is powerful.

The self is precious because it is fragile.

The Body Inside the Self

We often speak as if the self lives inside the body like a person inside a house.

But the self is not merely inside the body. The body is inside the self.

Hunger changes the self. Pain changes the self. Fever changes the self. Sleep changes the self. Hormones change the self. Fatigue changes the self. Aging changes the self.

A person who has not slept for two nights is not the same in patience, judgment, concentration, and emotional control. A person in severe pain may not speak with the same warmth. A person with low blood sugar may become shaky, anxious, confused, or irritable. A person with thyroid disease, infection, menopause, chronic illness, or medication side effects may feel different inside themselves.

These are not failures of the soul. They are body-brain realities.

The brain does not create the self in isolation. It is constantly receiving signals from the body: heartbeat, breathing, gut sensations, pain, temperature, muscle tension, hunger, thirst, fatigue, nausea, and many other internal states.

This may sound technical, but the everyday meaning is simple: how the body feels affects who we feel we are.

A calm body can make the world seem manageable. An anxious body can make the world seem threatening. A tired body can make life seem heavy. A healthy body can make the future seem open. A sick body can shrink the self into pain, fear, and survival.

This is why illness can change identity. A person diagnosed with cancer, heart disease, stroke, chronic pain, or dementia may begin to ask, "Who am I now?" The disease does not merely affect organs. It affects the story of the self.

Religion often separates body and soul. It treats the body as temporary and the soul as the true person. But human life does not divide so neatly.

The body is not just a shell. The body helps create the mind. The body helps create emotion. The body helps create the self.

We are not souls wearing bodies.

We are embodied selves.

Emotion and the Shape of Personality

We often imagine personality as something fixed.

"She is cheerful." "He is serious." "She is kind." "He is short-tempered." "She is brave." "He is anxious."

But personality is not a soul-stamp pressed into us from another world. It is shaped by temperament, brain development, genes, family, culture, trauma, learning, habits, illness, and choice.

Emotion plays a central role. A person who grew up in fear may carry fear into adulthood. A person repeatedly shamed may become defensive or withdrawn. A person loved with stability may develop trust more easily. A person who suffered betrayal may become guarded. A person who experienced kindness may become more open to giving it.

None of this requires a soul. It requires a living brain shaped by life.

Emotion colors memory. It guides attention. It shapes decisions. It influences what we notice, what we avoid, whom we trust, what we fear, and what we hope for.

This is why two people can live through the same event and become different in different ways. One becomes bitter. Another becomes compassionate. One becomes fearful. Another becomes determined. One closes down. Another opens up.

The self is not passively stored inside us. It is shaped through interaction with the world.

This view is more compassionate than the soul-based view. If personality were simply the expression of an inner soul, then weakness, fear, anger, addiction, or cruelty might be seen as spiritual failure alone. But when we understand the role of brain, trauma, habit, and environment, we ask deeper questions.

What pain shaped this person? What fear trained this response? What illness changed this behavior? What childhood created this defense? What social conditions encouraged this cruelty? What support could help this person change?

This does not erase responsibility. But it gives responsibility a human context. A person is not excused from harm because of background, but neither should we pretend that character forms in a vacuum.

The self is shaped. And because it is shaped, it can sometimes be reshaped.

That is one of the hopeful messages of a brain-based view. If the self were an eternal fixed soul, change would be mysterious. But if the self is a living pattern, then growth is possible.

Education can change us. Therapy can change us. Love can change us. Illness can change us. Grief can change us. Reading can change us. Science can change us. Honest doubt can change us.

A person can leave an old belief, reject an old prejudice, repair an old habit, soften an old anger, or build a new moral life.

The self is not a prison.

It is a process.

Other People Help Hold Us Together

No one creates a self alone.

From the beginning, other people help us become who we are. A baby does not know its name. Others give it a name. A child does not know its story. Others tell it: where it was born, who loved it, what happened, what kind of child it was.

A young person learns identity through family, language, school, religion, nationality, caste, class, gender expectations, praise, criticism, and social belonging. We learn who we are partly through how others see us.

"You are smart." "You are useless." "You are a good girl." "You are sinful." "You are brave." "You are not like us." "You are chosen." "You are impure." "You are loved." "You are not enough."

These words can enter the self and live there for decades.

Religion has often understood this power and used it. It tells children who they are before they can think for themselves. It tells them they are born sinful, chosen, impure, blessed, cursed, obedient, fallen, saved, or damned. These labels can become part of identity long before reason is mature enough to examine them.

A child raised to fear hell may carry that fear into adulthood, even after doubt begins. A girl raised to believe she is spiritually inferior may struggle with dignity. A person born into a caste system may be told that inequality is sacred order. A young person taught that doubt is sin may fear honest thought.

This shows something important: the self is social.

It is built not only by neurons and memories, but also by language, authority, family, culture, and power.

That is why liberation can feel like becoming oneself for the first time. When a person leaves a harmful belief system, rejects inherited prejudice, or stops seeing himself through religious fear, it is not merely a change of opinion. It can be a reconstruction of the self.

The person may say, “I finally feel free.”

What has happened?

A story has changed. The self has been revised.

This again makes better sense if the self is a living process rather than a fixed soul. Other people can wound the self. Other people can heal the self. Other people can help us remember who we are.

This is seen beautifully in dementia care. When memory weakens, family members, photographs, songs, familiar routines, touch, and loving voices can help hold parts of identity together. The person may not remember every fact, but relationship still matters. A song from youth may awaken emotion. A familiar poem or lullaby may return. A hand held gently may calm fear.

The self is not locked inside one isolated soul. It lives partly in connection.

We are held together by memory, body, brain, and one another.

The Self as a Story the Brain Keeps Revising

The title of this chapter says that the self is a living story.

This does not mean the self is fiction.

A story can be true. A family has a story. A marriage has a story. A nation has a story. A life has a story. These stories are not imaginary merely because they are organized by memory and meaning.

The self is a story in this sense: it is the brain's ongoing attempt to organize experience into a coherent "me." It connects the child you were with the adult you are. It links memory with purpose. It explains your wounds. It gives meaning to your choices. It turns events into identity.

But like all stories, it can change.

A painful childhood may first be remembered as shame. Later, it may be understood as survival. A failed business may first feel like humiliation. Later, it may become a chapter of learning. A lost faith may first feel like emptiness. Later, it may become liberation. A life of obedience may later be reinterpreted as a life lived under fear.

A person who once believed, “God made me,” may later say, “Nature made me, my family shaped me, my brain carries me, and my choices still matter.”

That is not the death of the self.

It is the growth of the self.

This is why the soul is not needed. The soul is too rigid an idea for the richness of human identity. It imagines a hidden fixed essence. But real people are more fluid, more wounded, more creative, more social, more biological, and more changeable than that.

We are not statues with souls inside.

We are living histories.

Why a Changing Self Does Not Need an Unchanging Soul

Religion often argues that because we feel like continuous persons, there must be an unchanging soul underneath the changes.

But continuity does not require an unchanging substance.

A city can have continuity though buildings, people, roads, and customs change over time. A friendship can have continuity though both people grow older and change. A body has continuity though cells are replaced, tissues remodel, and appearance changes. A life story can have continuity though new chapters are added, old chapters are reinterpreted, and some pages are lost.

Continuity can come from pattern, relationship, memory, structure, and process.

The self can be continuous in the same way, not because an immortal soul remains untouched, but because the brain and body maintain enough pattern from moment to moment, because memory links past and present, because other people recognize us, because language gives us a name, because habits persist, because the body continues, and because the story keeps being revised rather than completely erased.

This makes sense of real life. It explains why we can change and still feel continuous. It explains why illness can disrupt identity. It explains why memory matters. It explains why relationships matter. It explains why childhood matters. It explains why trauma can reshape the self. It explains why recovery and growth are possible.

The soul explains less. It simply says, “There is a hidden essence.”

But it does not show how memory works. It does not explain dementia. It does not explain why personality changes with brain injury. It does not explain how body signals shape emotion. It does not explain why childhood wounds enter adult identity. It does not explain why social rejection can damage the self or why love can help rebuild it.

The brain-based, body-based, relationship-based view explains more.

And it does so without removing dignity.

The self is not less precious because it is a process. A life is not less meaningful because it is built through time. A person is not less real because the person changes.

The self is not a hidden object.

It is a living history.

The Soul Was an Answer to the Feeling of "I"

The feeling of "I" is one of the most intimate experiences we have.

I am here. I am thinking. I remember. I choose. I suffer. I love. I will die.

This feeling is so deep that it is easy to make a leap. Because we feel like an inner "I," we assume there must be an invisible thing inside us that owns experience.

But feeling is not final proof.

The brain can create powerful experiences that do not match external reality. Dreams feel real while they happen. Phantom limb pain can be felt in a limb that is no longer there. Optical illusions can make us see what is not present. Anxiety can make danger feel near when none exists. Depression can make the future feel hopeless even when change is possible.

The feeling of a permanent inner "I" may also be a construction. It may be the brain's way of organizing experience for survival.

A living organism must distinguish self from world. It must protect its body. It must remember dangers. It must recognize allies. It must plan for the future. It must learn from past actions. It must maintain enough continuity to function.

The feeling of self helps do all this. It is useful. It is powerful. It is real as experience.

But usefulness does not prove immortality.

The self may be nature's way of organizing a life, not heaven's signature inside the body.

What This Means for Human Dignity

Some people fear that without a soul, the person becomes "only biology."

But that word "only" is doing too much damage.

A human being is biology, but not "only" biology in any empty sense. A brain is biology, but it can produce mathematics, music, memory, grief, justice, medicine, literature, and love. The heart is biology, but a mother holding a sick child is not "only" biology in any shallow sense. A tear is biology, but sorrow is not meaningless. A kiss is biology, but love is not fake. A moral choice is brain activity, but that does not make it trivial.

The mistake is thinking that natural means worthless.

This is one of religion's most harmful habits. It often teaches people that if something is not supernatural, it is not sacred, not meaningful, not dignified, not worthy.

But why should that be true?

A forest is natural. It is still beautiful. A child is natural. A child is still precious. The brain is natural. The mind is still astonishing. The self is natural. The self is still worthy of respect.

Human dignity does not need an immortal soul. It comes from the reality of conscious life: the ability to feel pain, experience joy, form bonds, remember, hope, fear, love, suffer, and flourish.

That is enough.

A person matters because a person can be harmed. A person matters because a person can suffer. A person matters because a person can love and be loved. A person matters because life is experienced from the inside.

No soul is required for this.

The Freedom of a Changing Self

If the self is not fixed, then we are not trapped forever inside an unchanging essence.

This can be frightening, but it is also hopeful.

We can change. We can learn. We can revise old beliefs. We can leave inherited fear. We can outgrow prejudice. We can heal from shame. We can become more compassionate. We can become more honest. We can become more free.

Many people live for years inside a religious identity they did not choose. They are told who they are before they know how to question. They inherit a God, a scripture, a community, a fear of hell, rules about purity, ideas about gender, stories about sin, and a picture of the soul.

Later, some begin to ask questions. At first, doubt may feel like danger. Then it may feel like grief. Then it may feel like betrayal. Then, slowly, it may feel like air.

The self changes.

A person who once said, "I am a sinner being watched by God," may one day say, "I am a human being responsible for my life." A person who once feared divine punishment may one day feel moral responsibility without fear. A person who once believed life was a test for heaven may one day see life as precious because it is brief.

This is not moral collapse. It is growth.

It is the self revising its deepest story.

Science-based atheism is not only a change in belief. For many people, it is a change in identity. It moves the self from fear to responsibility, from obedience to understanding, from supernatural dependence to human maturity.

This is why the question of the self matters so much.

If we are fixed souls, then our deepest identity belongs to theology.

If we are living selves created through brain, body, memory, and relationship, then our identity remains open to truth.

And truth can change us.

Chapter Summary

This chapter examined the self: the inner feeling of being one continuous "I" through time.

Religion often explains this feeling by claiming that an immortal soul lives inside the body. But a brain-based view gives a better explanation. The self is not a fixed supernatural object. It is a living process created by memory, body signals, emotion, personality, language, relationships, culture, and time.

Memory helps create the story of who we are. The body shapes the self through hunger, pain, fatigue, illness, hormones, and internal signals. Emotions and relationships shape personality and identity. Other people help form, wound, and sometimes heal the self.

The self is real, but it is not unchanging. It can grow, break, heal, revise itself, and reinterpret the past.

This does not make human dignity weaker. It makes human beings more understandable and more precious. We matter not because we contain immortal souls, but because we are conscious, feeling, vulnerable beings who can suffer, love, remember, hope, and change.

Key Takeaways

The feeling of a permanent inner "I" is real as an experience, but it does not prove an immortal soul.

Memory is one of the main threads holding personal identity together.

The body is not merely a container for the self. Body signals help shape emotion, awareness, and identity.

Relationships, language, culture, and social roles help create the story of who we are.

A changing self does not need an unchanging soul. Continuity can come from memory, pattern, body, relationship, and story.

Human dignity does not require a soul. It rests on the reality of conscious life, suffering, love, and moral responsibility.

Transition to the Next Chapter

The self feels deeply real. But behind the self lies an even deeper mystery: consciousness itself.

How does the brain create experience? How does light become color, sound become music, touch become tenderness, and memory become identity?

Religion often points to consciousness and says, “Here is where the soul must be.”

The next chapter will examine that claim.

Consciousness is astonishing. It is still not fully understood. But mystery is not the same as magic.

And an unanswered question is not evidence for God.

Sources

Antonio Damasio, *Self Comes to Mind: Constructing the Conscious Brain.*

Anil Seth, *Being You: A New Science of Consciousness.*

Daniel L. Schacter, *The Seven Sins of Memory.*

Oliver Sacks, *The Man Who Mistook His Wife for a Hat.*

H. J. Markowitsch and A. Staniloiu, “Memory and Self: Neuroscientific Landscapes,” *Frontiers in Psychology*, 2013.

Arnaud D’Argembeau, “Memory, Future Thinking, and the Self,” *Memory & Cognition*, 2021.

A. Monti et al., “The Inside of Me: Interoceptive Constraints on the Concept of Self,” *Neuroscience & Biobehavioral Reviews*, 2021.

C. Stramba-Badiale et al., "Autobiographical Memory in Alzheimer's Disease: A Systematic Review," 2025.

A. D'Argembeau et al., "Brains Creating Stories of Selves: The Neural Basis of Autobiographical Reasoning," *Social Cognitive and Affective Neuroscience*, 2013.

Chapter 4: Consciousness Without Magic

Why Mystery Is Not Evidence for God

Consciousness is astonishing.

The brain turns light into color. It turns vibration into music. It turns pressure on the skin into touch. It turns memory into identity. It turns injury into pain. It turns another person's face into love, grief, trust, fear, or longing.

Every morning, when you wake, the universe appears again inside your awareness. The room becomes familiar. The body becomes yours. The past returns. The day begins. The private feeling of "I am here" rises again from sleep.

This is one of the deepest mysteries in nature.

A stone does not know that it exists. A chair does not feel sunlight. A river does not wonder where it came from. But a human brain can look at the night sky and ask why anything exists at all.

No wonder religion claimed consciousness for the soul.

For thousands of years, human beings looked at thought, memory, imagination, love, fear, and awareness, and felt that something so intimate could not be merely physical. The body seemed like matter. Consciousness seemed like spirit. The flesh could be touched, but the inner world could not. So people imagined an invisible self inside the visible body.

Even today, many religious people say, "Science may explain the brain, but it cannot explain consciousness. Therefore, there must be a soul."

This argument feels powerful because consciousness is so close to us. We do not merely observe it. We live inside it. Everything we know, we know through consciousness. Even doubt, reason, science, prayer, fear, and love all appear inside awareness.

But the emotional force of an argument is not the same as evidence.

Consciousness is mysterious, but mystery is not magic. A hard question is not proof of God. An unanswered question is not evidence for the soul.

The history of knowledge teaches us to be careful. Many things once seemed too mysterious for natural explanation. Lightning, disease, epilepsy, dreams, birth, mental illness, eclipses, and the origin of species were all placed under supernatural control. Then science advanced. The mystery did not vanish, but the supernatural explanation did.

Consciousness may be the deepest remaining mystery of the mind.

But it still belongs to nature.

The Wonder of Being Aware

Pause for a moment and look at an ordinary object: a cup, a book, a hand, a chair, a window. You do not experience raw light waves. You experience shape, color, location, meaning, and familiarity. You do not merely receive information. You experience a world.

Now listen. A sound may become a word, a warning, a song, a laugh, a memory, or a familiar voice. The ear receives vibration, but the brain creates meaning.

Now touch your own hand. You feel pressure, warmth, texture, and ownership. It is not just a hand. It is your hand.

This is consciousness: the presence of experience. It is what makes pain painful, love felt, music moving, food enjoyable, injustice unbearable, and death frightening.

Without consciousness, there would still be matter, stars, oceans, cells, and chemical reactions. But there would be no experienced world. There would be no joy, no sorrow, no beauty, no hope, no fear, no meaning.

This is why consciousness feels sacred to many people.

But we must be careful with that word. Something can be deeply valuable without being supernatural. A newborn's cry is natural, but it is still precious. A forest is natural, but it is still beautiful. A heart rhythm is biological, but it can still mean life.

Consciousness may be natural and still be wondrous.

In fact, the wonder becomes deeper when we stop treating consciousness as a magical substance and begin asking how the brain creates it.

How does the brain combine signals from eyes, ears, skin, body, memory, and emotion into one experienced world? How does it create the sense of being located inside a body? How does it distinguish self from non-self? How does it allow attention, choice, imagination, and reflection?

These are scientific questions. They are difficult questions. But they are not meaningless questions.

The very fact that scientists can study consciousness through brain injury, anesthesia, sleep, imaging, electrical activity, behavior, and reports of experience shows that consciousness is not beyond investigation.

Religion claims consciousness.

Science studies consciousness.

Studying is better than claiming.

Why Consciousness Feels So Mysterious

Consciousness feels mysterious partly because we experience it from the inside.

You can see another person's body. You can hear their words. You can observe their actions. You can measure their pulse, scan their brain, record electrical activity, and watch their behavior. But you cannot directly enter their experience.

You can know that another person reports pain, but you cannot feel their pain exactly as they feel it. You can know that another person sees red, but you cannot step inside their redness. You can know that a child is afraid, but you cannot fully inhabit that child's fear.

This private quality of experience makes consciousness different from many other scientific subjects. A bone can be X-rayed. Blood can be tested. A heart valve can be seen on ultrasound. But the feeling of pain, the taste of mango, the grief of losing a parent, or the beauty of a song cannot be placed on a table.

This does not mean consciousness is supernatural. It means consciousness is subjective.

Subjective does not mean unreal. Pain is subjective, but it is real. Grief is subjective, but it is real. Love is subjective, but it is real. Color experience is subjective, but it is real. A headache cannot be handed to another person, but no sensible person says headaches do not exist.

The difficulty is explaining how subjective experience arises from physical processes in the brain.

Philosophers sometimes call this the "hard problem" of consciousness. Why should brain activity feel like anything from the inside? Why should neurons firing in darkness create the redness of red, the bitterness of coffee, the ache of loneliness, or the warmth of affection?

This is indeed difficult. But difficulty should not be confused with divinity.

In the past, life itself seemed impossible to explain naturally. People thought living things must contain a special vital force. Then biology advanced. We learned about cells, metabolism, genes, evolution, development, and molecular machinery. Life remained astonishing, but the need for a supernatural life-force faded.

The same may happen with consciousness.

At present, we do not have a complete explanation. But we have many clues, and those clues point toward the brain.

Consciousness changes with sleep, anesthesia, seizures, brain injury, drugs, oxygen failure, dementia, development, attention, and arousal. This is exactly what we would expect if consciousness depends on brain processes.

It is not what we would expect if consciousness were produced by an independent soul.

The Brain Builds Experience

The brain does not passively receive the world like a camera. It builds experience.

This is easy to miss because consciousness feels immediate. When you open your eyes, the world simply seems to be there. But between light entering the eye and your experience of a meaningful world, the brain performs extraordinary work.

It detects edges, color, movement, depth, contrast, faces, objects, threats, patterns, and meaning. It uses memory to interpret what is seen. It uses expectation to predict what is likely. It uses emotion to decide what matters. It uses attention to select what becomes prominent.

You do not experience everything around you equally. You experience what the brain selects, organizes, and presents.

This is why illusion is possible. A magician can guide attention and make you miss what is in front of your eyes. A visual illusion can make two equal lines look unequal. A shadow can be mistaken for a person. A dream can feel real while you are inside it. A hallucination can create an experience without an external object. A phantom limb can produce pain in a limb that is no longer there.

These examples show that experience is not a simple copy of the outside world. It is a construction.

Again, construction does not mean fake. A map is constructed, but it can still guide us. A melody is constructed from notes, but it can still move us. Conscious experience is constructed by the brain, but it is still real experience.

The brain also builds the feeling of a unified world. Your visual experience, body position, sounds, memories, emotions, and thoughts are usually woven together into one scene. You do not experience life as scattered signals. You experience an organized world centered around a self.

This unity can be disrupted.

Brain injury can affect awareness of one side of space. Some patients with neglect may ignore the left side of the world after damage to certain brain regions. Other patients may deny paralysis. Some may fail to recognize familiar faces. Others may lose the emotional feeling that normally comes with recognition.

These disorders are powerful because they show that ordinary consciousness is not simple. It is assembled. When assembly is disturbed, the world of experience changes.

A soul theory gives us little help here.

A brain-based theory does.

What Science Is Trying to Explain

Science has not produced one final theory of consciousness. That should be admitted honestly.

But the absence of a final answer does not mean there is no progress. There are several major scientific theories, and they all try to explain consciousness through natural processes rather than supernatural souls.

One theory is called Global Neuronal Workspace Theory. In simple language, it suggests that information becomes conscious when the brain makes it widely available to many systems. Something may be processed unconsciously, but when it becomes broadly available for attention, memory, planning, and report, it enters consciousness.

Another theory is called Integrated Information Theory. In simple language, it suggests that consciousness depends on how deeply information is unified inside a system. Conscious experience is not just scattered signals. It feels like one whole experience.

Another approach, often connected with predictive processing, says the brain is constantly predicting the world. The brain does not simply wait for information. It guesses, checks, corrects, and updates. Conscious experience may be part of this active process of building a useful model of reality.

There are other theories too. Scientists disagree. Some theories may be partly right. Some may be wrong. Some may need major revision.

This is normal. Science does not become weak because it has competing theories. It becomes stronger by testing them.

Religion often presents certainty first and evidence later. Science does the opposite. Science begins with uncertainty, gathers evidence, tests predictions, corrects errors, and accepts that some cherished ideas may fail.

That humility is not weakness. It is intellectual honesty.

A religious believer may look at scientific disagreement and say, “See, science does not know. Therefore, God.”

But that is a poor argument.

When doctors debate the best treatment for a disease, that does not mean the disease is caused by demons. When physicists debate the interpretation of quantum mechanics, that does not mean angels move electrons. When neuroscientists debate theories of consciousness, that does not mean a soul is hiding in the gap.

Debate is not defeat.

Uncertainty is not permission to invent certainty.

The God of the Gaps Inside the Mind

The “God of the gaps” argument is one of the oldest habits of religion.

Where knowledge is incomplete, God is placed. People did not understand thunder, so thunder belonged to gods. People did not understand disease, so disease belonged to spirits, curses, or divine punishment. People did not understand life’s complexity, so life belonged to a designer. People did not understand consciousness, so consciousness belonged to the soul.

The problem is that the gaps keep shrinking.

Every time knowledge grows, God retreats.

This does not prove that every mystery has already been solved. It simply shows that mystery has often been a poor foundation for supernatural belief.

Consciousness is now one of the most attractive remaining gaps. Religious apologists often say, “Science cannot explain why there is something it is like to be you. Therefore, consciousness must come from a soul or from God.”

But this argument makes a leap. It moves from “science has not fully explained consciousness” to “therefore my religion is true.”

That leap is not justified.

Even if neuroscience has not solved consciousness, that does not prove Hinduism, Christianity, Islam, Judaism, or any other religion. It does not prove heaven, hell, karma, reincarnation, angels, demons, judgment, or divine command. It does not prove that consciousness survives death. It does not prove that a soul exists.

At most, it proves that consciousness is difficult.

And we already know that.

A mystery does not belong automatically to religion. It belongs to inquiry. If a person says, “I do not know exactly how consciousness arises,” that is honest. If a person says, “I do not know, therefore God,” that is not knowledge.

That is surrender.

The mind deserves better than surrender.

Disorders of Consciousness and the Brain

Some of the clearest evidence that consciousness depends on the brain comes from disorders of consciousness.

Severe brain injury, stroke, oxygen deprivation, infection, and other conditions can produce coma, unresponsive wakefulness, minimally conscious states, and other disorders. These conditions are medically complex and emotionally devastating.

A family may stand at a bedside and wonder: Is he aware? Can she hear us? Is there pain? Is there hope? Is the person still there?

Modern medicine approaches these questions not by searching for a soul, but by examining the brain: reflexes, eye movements, sleep-wake cycles, responses to command, brain imaging, electrical activity, and signs of awareness.

This is not because doctors are cold. It is because consciousness depends on brain function.

This has real human importance. Some patients who appear unresponsive may retain signs of hidden awareness detectable through careful testing. That possibility matters deeply for care, communication, prognosis, and ethical decisions. It also reinforces the central point: if we want to understand consciousness, we study the brain.

We do not weigh the soul. We do not scan heaven. We do not measure spirit. We examine brain systems.

Again, the brain-based view does not remove compassion. It increases it. It helps us recognize that awareness can be hidden, damaged, partial, fluctuating, or recovering. It encourages careful diagnosis and humility in medical decisions.

A soul-based explanation does not guide this work.

Neuroscience does.

The Soul Cannot Explain the Fragility of Awareness

Consciousness is fragile.

This fragility is one of the strongest arguments against the soul.

If consciousness comes from an immortal soul, why does it disappear in deep sleep? Why is it altered by fever? Why is it distorted by delirium? Why can a seizure interrupt it? Why can anesthesia silence it? Why can oxygen failure erase it? Why can brain injury reduce it? Why can dementia gradually change the experienced world?

A soul should not be so vulnerable to blood flow, chemicals, electrical storms, infection, and damaged tissue.

But a brain process would be.

Some people use a radio analogy to defend the soul. They say the brain is like a receiver, and the soul or consciousness is like a signal. If the radio is damaged, the music is distorted, but the signal still exists somewhere else.

At first, this sounds plausible. But the analogy fails.

When the brain is damaged, it is not merely the "sound quality" of consciousness that changes. The contents of consciousness can change. Memory can disappear. Language can break. Recognition can vanish. Personality can shift. Awareness can fade. The sense of self can fragment. Moral judgment can weaken. Hallucinations can appear.

This is not like a radio receiving a perfect signal poorly. It is more like the brain is generating the experience.

Damage the generator, and the generated process changes.

The soul theory adds an invisible entity but explains less. The brain theory explains why consciousness rises and falls with the state of the nervous system.

That is why it is stronger.

Wonder Without Superstition

Some people fear that explaining consciousness through the brain makes human life dull.

They imagine that if consciousness is natural, then beauty disappears. Love becomes chemicals. Music becomes vibrations. Thought becomes electricity. The self becomes machinery.

But this is a failure of imagination.

Natural does not mean empty. A sunset is natural. It is still beautiful. A child's laughter is natural. It is still joyful. A heartbeat is biological. It is still precious. A poem is made of words. It can still change a life. A brain is made of cells. It can still create wonder.

The idea that only the supernatural can be meaningful is one of religion's deepest mistakes.

Science does not reduce consciousness to "nothing but neurons" in the cheap sense. It reveals that neurons, networks, bodies, memories, emotions, and environments can produce something astonishing: a lived world.

That should not make us feel smaller. It should make nature feel larger.

The universe produced stars, planets, oceans, cells, forests, animals, and eventually brains capable of knowing that the universe exists.

Matter became aware.

That is more astonishing than a soul placed inside the body by divine command.

A soul is an explanation that stops the question. The brain is a mystery that invites deeper understanding.

Consciousness and Death

The deepest emotional reason people connect consciousness with the soul is death.

Each of us experiences consciousness as the center of life. To imagine its end is frightening. We cannot easily imagine nonexistence because imagination itself requires existence. So the mind rebels. It says, "I cannot simply end. Something of me must continue."

Religion answers that fear with the soul.

It says consciousness survives the death of the body. It says the self continues in heaven, hell, rebirth, spirit realm, or some other invisible world.

But again, desire is not evidence.

The fact that we want consciousness to continue does not prove that it does.

Everything we know about ordinary consciousness shows dependence on the living brain. If sleep alters consciousness, if anesthesia interrupts it, if brain injury damages it, if dementia changes it, if oxygen failure removes it, then it is difficult to argue that consciousness floats free after the brain permanently dies.

The evidence points in another direction: when the brain can no longer support consciousness, consciousness ends.

This is painful. But pain does not make it false.

Some truths are hard because they ask us to love life without guarantees. They ask us to value people without imagining they are immortal. They ask us to treat the present moment as precious because it is not promised again.

A mortal consciousness is not worthless. It may be more precious because it is temporary.

A candle flame matters though it does not burn forever. A song matters though it ends. A life matters though it is finite.

Consciousness does not need immortality to be meaningful. It needs experience, value, relationship, and time.

For now, we have those.

That is why this life matters.

Mystery Is a Beginning, Not an Answer

There is nothing wrong with saying, "I do not know."

Those may be the most honest words in human language.

Science begins there.

I do not know how lightning works. I do not know what causes infection. I do not know why species change. I do not know how memory forms. I do not know how consciousness arises.

But science does not stop at "I do not know." It asks, observes, tests, compares, revises, and learns.

Religion too often turns "I do not know" into "God did it." That may feel satisfying, but it is not explanation. It is closure without understanding.

Consciousness deserves better.

The right response to mystery is not worship of ignorance. The right response is disciplined wonder: wonder that asks questions, respects evidence, admits uncertainty, and does not invent answers merely because the truth is incomplete.

The brain is still not fully understood. Consciousness is still not fully explained. The subjective feeling of experience remains one of the hardest problems in science and philosophy.

But we know enough to say this: consciousness depends deeply on the brain.

That does not solve everything. But it removes the need for the soul as an explanation.

The soul was once used to fill the mystery of awareness. But the more we learn, the more consciousness looks like a natural process of living brains, not a supernatural substance placed inside bodies.

We do not need to deny the wonder.

We only need to deny the magic.

Chapter Summary

Consciousness is one of the deepest mysteries in nature. It is the presence of experience: the feeling of seeing, hearing, touching, remembering, loving, suffering, and being a self.

Religion has often claimed that consciousness must come from the soul because it feels too private, too rich, and too mysterious to be physical. But mystery is not evidence for God.

Modern science studies consciousness through brain activity, sleep, anesthesia, brain injury, disorders of consciousness, attention, perception, and competing theories. These theories disagree in important ways, but they all treat consciousness as a natural phenomenon open to study.

Brain injury, anesthesia, sleep, seizures, dementia, oxygen failure, and disorders of consciousness all show that awareness depends deeply on brain function. This is difficult to explain if consciousness comes from an independent soul. It is exactly what we should expect if consciousness is created by the living brain.

Consciousness remains mysterious, but it does not require magic.

Key Takeaways

Consciousness is astonishing, but astonishment is not proof of the supernatural.

The private nature of experience makes consciousness difficult to explain, but difficulty is not evidence for God.

Modern consciousness research is active, empirical, and testable, even though no final theory has won.

Disorders of consciousness after brain injury show that awareness depends on brain systems.

The soul does not explain why consciousness changes with sleep, anesthesia, drugs, oxygen, seizures, dementia, and brain injury.

A natural view of consciousness does not destroy wonder. It deepens wonder by showing what living brains can do.

Transition to the Next Chapter

Consciousness feels continuous while we are awake, but every day it changes.

In sleep, the world fades. In dreams, the brain creates private realities. Under anesthesia, awareness can disappear and then return. In coma, consciousness can be damaged, hidden, or lost.

If the soul is the source of awareness, these changes are hard to explain. But if consciousness depends on the brain, they make sense.

The next chapter will look at these altered states of awareness and ask a direct question:

If the soul is always present, why does the self disappear when the brain changes state?

Sources

Anil Seth, *Being You: A New Science of Consciousness.*

Stanislas Dehaene, *Consciousness and the Brain.*

Antonio Damasio, *Self Comes to Mind: Constructing the Conscious Brain.*

Christof Koch, *The Feeling of Life Itself.*

David Chalmers, *The Conscious Mind.*

Cogitate Consortium et al., "Adversarial Testing of Global Neuronal Workspace and Integrated Information Theories of Consciousness," *Nature*, 2025.

G. Friedman et al., "The Current of Consciousness: Neural Correlates and Clinical Aspects of Consciousness," 2023.

J. T. Giacino et al., "Comprehensive Systematic Review Update Summary: Disorders of Consciousness," *Neurology*, 2018.

Y. G. Bodien et al., "Disorders of Consciousness Diagnosis, Interventions, and Outcomes," 2025.

B. L. Edlow et al., "Recovery from Disorders of Consciousness," *Nature Reviews Neurology*, 2020.

Chapter 5: When the Self Disappears

Sleep, Dreams, Anesthesia, Coma, and the Fragility of Awareness

Every night, the world disappears.

You close your eyes. The room fades. The body becomes still. The familiar world of faces, duties, worries, plans, and memories begins to loosen. The firm sense of "I am here" grows dim.

Then, for a time, the self is gone.

You do not watch the clock. You do not hear every sound. You do not feel the movement of Earth through space. You do not think, "I am unconscious now." You simply disappear from your own experience.

Then morning comes. The eyes open. The room returns. The body becomes yours again. Memory reconnects. Your name returns. The past returns. The unfinished tasks return. The self rises again.

This ordinary daily event is one of the strongest arguments against the soul. If consciousness belongs to an immortal soul, why does it fade every night? If the soul is the true center of awareness, why does awareness depend so strongly on the state of the brain?

Sleep is not death. But sleep gives us a daily lesson: the conscious self is not constant. It rises and falls. It can be dimmed, altered, interrupted, hidden, or lost.

Dreams make the mystery even stranger. While the body lies in bed, the brain creates private worlds. Dead people return. Childhood houses reappear. Fear becomes a chase. Desire becomes a story. Shame becomes a scene. A person can fly, fall, run, speak, cry, or meet someone long gone.

Then the dream ends. The world that felt real dissolves.

Under anesthesia, the disappearance is deeper. A patient lies on an operating table. Medicine enters the body. Awareness fades. Pain disappears. Memory stops. Then, after surgery, consciousness returns as if someone opened a curtain.

In coma and disorders of consciousness, the question becomes even more painful. A person may lie still, eyes closed or open, while family members ask: Is he there? Can she hear us? Is there awareness? Is there suffering? Is there hope?

In all these states, one truth repeats. Consciousness is fragile.

The self is not an unchanging light carried by an immortal soul. It behaves more like a flame that depends on conditions. Change the conditions, and the flame changes. Remove the conditions, and the flame goes out.

This is not a small observation. It reaches into the center of what we are.

The Nightly Disappearance of the World

Sleep is so common that we forget how strange it is.

Every day, every human being must surrender consciousness for several hours. The proud mind, the planning mind, the religious mind, the scientific mind, the loving mind, the anxious mind, the ambitious mind, the doubting mind, all must sleep.

No belief system prevents this. No prayer keeps ordinary consciousness fully awake forever. No soul remains continuously alert while the brain sleeps.

Sleep is not simply rest for the body. It is an altered state of the brain and mind. During sleep, awareness of the external world changes. Responsiveness decreases. The brain cycles through different stages. Memory, emotion, and bodily regulation are affected.

This is important. If the soul is the true seat of consciousness, why should consciousness require sleep? Why should awareness fade when the brain enters sleep states? Why should memory, emotion, and thinking change when sleep is poor? Why should a person become irritable, confused, depressed, forgetful, or slow after sleep deprivation?

The answer is not that the soul is tired. The answer is that the brain is a biological organ. Sleep changes the brain, and the changed brain changes the mind.

We see this in ordinary life. A person who sleeps well may wake with patience, clarity, and emotional balance. The same person after several nights of poor sleep may become short-tempered, anxious, unfocused, forgetful, and more reactive. Nothing supernatural has changed. No demon has entered. No soul has weakened.

The brain is sleep-deprived. The mind follows.

Sleep shows us that consciousness is not a permanent spiritual flame burning independently of the body. It is a living process that depends on the state of the nervous system.

Every night, the self does not leave the body. The brain changes state, and the conscious world fades.

Dreams and the Brain's Private Theater

Dreams are among the most powerful reasons human beings invented the soul.

In dreams, the body lies still, but the mind travels. A person may dream of a childhood home demolished long ago. A dead mother may speak again. A friend may appear young again. A person may run from danger, feel shame, taste food, hear music, or wake with tears.

To an ancient person, dreams must have seemed like evidence that an inner self could leave the body.

Even today, dreams can feel spiritually meaningful. A grieving man dreams of his dead wife and feels she visited him. A mother dreams of a lost child and feels comfort. A person dreams of an ancestor and feels guided.

These experiences are emotionally real.

But emotional reality is not the same as supernatural reality.

Modern neuroscience studies dreams as experiences generated by the sleeping brain. Dreams may draw from memory, emotion, fear, recent experience, old wounds, grief, desire, and imagination. They can feel strange because the dreaming brain does not organize reality in the same way the waking brain does.

This matters because dreams show that the brain can create a world without the outside world. The eyes are closed, yet images appear. The ears receive little from the room, yet voices are heard. The body lies still, yet the dreamer runs. The dead are absent, yet they speak. The past is gone, yet it returns.

This does not show that the soul travels. It shows that the brain can generate experience internally.

Dreams are not meaningless. They often carry emotional truth. A dream can comfort. A dream can disturb. A dream can reveal grief that waking life has hidden. A dream can bring back the dead in the theater of memory. A dream can remind us that the brain is not only a receiver of reality. It is also a creator of inner worlds.

But this is not evidence for the soul. It is evidence for the astonishing creativity of the brain.

The dead return in dreams because memory, love, and grief remain in the living brain. That is tender, powerful, and human. But it is not proof of another world.

Lucid Dreams and the Shifting Witness

Sometimes, a dreamer knows he is dreaming while the dream continues. This is called lucid dreaming.

A person may be inside a dream and suddenly think, "This is a dream." The dream world may continue, but the dreamer now has some awareness of the state he is in. The dream may even become partly controllable.

This is fascinating because it shows that self-awareness itself can vary during sleep. In ordinary dreaming, the dreamer is inside the story and accepts it. In lucid dreaming, part of the mind recognizes the story as a dream. In deep sleep, there may be little or no remembered experience. In waking life, the world is stable and shared with others.

These are not separate souls entering different realms. They are different brain states producing different forms of awareness.

The self is not one fixed lamp. It is more like a dimmer switch: sometimes bright, sometimes faint, sometimes distorted, sometimes absent, sometimes aware of itself, sometimes lost in its own creations.

This makes better sense in a brain-based view than in a soul-based view.

The soul is supposed to be the stable witness.

But the witness itself changes.

Anesthesia and the Chemical Silence of Consciousness

General anesthesia is one of the most powerful facts in modern medicine.

A patient comes into the operating room awake, anxious, and aware. The lights are bright. The monitors beep. The staff moves with practiced calm. The anesthesiologist gives medication.

Then awareness fades.

The patient does not merely become quiet. The patient's conscious world disappears.

No pain. No time. No operating room. No memory.

Later, the eyes open.

"Is it over?"

To the patient, surgery may feel like a missing piece of time. One moment, there was fear before the procedure. The next moment, there is recovery.

What happened to the self?

If consciousness belongs to a soul, why can a drug silence it?

This question should not be passed over lightly. Anesthesia is not a philosophical speculation. It happens every day in hospitals and surgical centers across the world. It is one of medicine's great achievements. It allows surgery without the full horror of pain and awareness. But it also reveals something profound about the mind.

Change the brain chemically, and consciousness changes.

This does not mean anesthesia is simple. Different anesthetic drugs work in different ways. The brain may not be completely "off." Some functions continue. The body is monitored carefully. Reflexes, breathing, blood pressure, and brain responses vary. There are rare and frightening cases of awareness under anesthesia, which show that consciousness during anesthesia is complex and must be taken seriously.

But the larger point remains. Consciousness can be profoundly changed by changing brain function.

The soul does not explain this well. If awareness were carried by an immortal soul, anesthesia should not erase the world. At most, it might prevent the body from moving or speaking. But anesthesia can remove conscious experience and memory. It can interrupt the person's experienced world.

The simplest explanation is that consciousness depends on the working brain. When the brain is placed into certain drug-induced states, the conscious self can disappear from experience. When the brain returns from those states, the conscious self can return.

No soul is needed.

Coma and the Injured Machinery of Awareness

Coma is one of the most heartbreaking conditions in medicine.

A person lies still. The eyes may remain closed. There is no normal waking response. The family waits. Machines may breathe for the patient. Tubes, monitors, and medications surround the bed. The room becomes a place of hope, fear, prayer, science, and uncertainty.

Families ask the questions any loving person would ask. Can he hear me? Does she know I am here? Is there pain? Will he wake up? Is she still there?

These questions should be approached with humility. Disorders of consciousness are complex. Some patients recover. Some do not. Some show minimal signs of awareness. Some appear unresponsive but may have hidden awareness detectable by advanced testing. Medical science continues to improve diagnosis, prognosis, and treatment.

But coma and related conditions point strongly to one fact: awareness depends on brain function.

Modern medicine does not evaluate coma by searching for the soul. It evaluates breathing, pupils, reflexes, eye movements, brain imaging, electrical activity, sleep-wake patterns, responses to voice, responses to pain, and signs of purposeful action.

This is not because doctors are cold. It is because the brain is where the evidence is.

If the soul were independently conscious, then coma should not be so dependent on brain injury, oxygen, swelling, bleeding, infection, sedation, and neural networks. The soul should remain awake, even if the body cannot respond.

Some religious people may say, “The soul is still there, but trapped.”

But that answer does not explain the details. It does not explain why different injuries produce different levels of awareness. It does not explain why some patients recover as brain function improves. It does not explain why electrical activity, imaging, sleep-wake cycles, reflexes, and responses to command matter. It does not guide treatment. It does not improve diagnosis.

Medicine studies the brain because the brain is where the evidence leads.

That is not cold materialism. It is care guided by reality.

The Painful Question: Is Anyone Still There?

In disorders of consciousness, the hardest question is often not scientific but emotional.

Is the person still there?

Families may speak, sing, touch, pray, and wait. A wife may hold the hand of a husband who cannot answer. A parent may sit beside a child after a car accident. A son may play his mother's favorite song. A daughter may whisper, "I am here."

These acts matter.

Even when awareness is uncertain, human tenderness is never wasted.

Science does not tell us to treat such patients as objects. On the contrary, it asks us to be careful, humble, and compassionate. Because awareness can be difficult to detect, clinicians must avoid careless assumptions. Some patients with severe brain injury may show signs of consciousness only through careful testing or advanced methods.

This is an important moral point. A brain-based view does not reduce the person. It protects the person by insisting on careful evidence.

It asks: what can this brain still do? What signs of awareness are present? What responses can be detected? What is the chance of recovery? What suffering must be prevented? What dignity must be preserved?

A soul-based view may offer comfort, but it does not answer these clinical questions. A brain-based view does.

The tragedy of coma does not prove the soul. It shows how fragile consciousness is when the brain is injured.

And because consciousness is fragile, care matters even more.

Why the Soul Cannot Explain the Fragility of Awareness

The soul is usually described as the deepest and most stable part of a person. The body may change, but the soul remains. The brain may age, but the soul remains. The world may wound us, but the soul remains.

This sounds comforting. But altered states of consciousness make the idea difficult to defend.

In deep sleep, awareness may disappear. In dreams, awareness becomes strange and unstable. In lucid dreams, part of the mind knows it is dreaming. Under anesthesia, consciousness can be chemically silenced. In coma, awareness can be lost or hidden after brain injury. In delirium, consciousness becomes confused and fluctuating. In seizures, consciousness can be interrupted. In dementia, the experienced world may slowly change.

If consciousness comes from the soul, why is it so dependent on the brain?

A soul should not need oxygen. A soul should not be confused by fever. A soul should not be silenced by anesthetic drugs. A soul should not dream because the sleeping brain is active. A soul should not lose awareness because brain networks are disrupted. A soul should not forget because memory systems are damaged.

But the mind does all these things because the mind is not an independent soul. It is a living process of the brain and body.

This is not an insult to consciousness. It is a more accurate understanding of it.

The flame is not insulted when we understand fire. The rainbow is not insulted when we understand light. The mind is not insulted when we understand the brain.

The Self Is More Like a Flame Than a Stone

Religion often imagines the soul as a thing: a hidden object, a spiritual substance, a permanent inner jewel, something that remains unchanged through childhood, age, sickness, sleep, and death.

But consciousness does not behave like a stone hidden inside the body. It behaves more like a flame.

A flame is real, but it depends on conditions. It needs fuel, oxygen, heat, and chemical reaction. Change the conditions, and the flame changes. Remove the conditions, and the flame ends.

The mind may be like that. It is real. It gives warmth, light, and meaning to a life. It can be bright, dim, steady, flickering, distorted, or gone. It depends on conditions:

blood flow, oxygen, glucose, sleep, brain networks, development, memory, body signals, chemical balance, and health.

This does not make the mind worthless. It makes it precious.

A fragile thing is not less valuable because it is fragile. Often, fragility increases value. A glass sculpture matters partly because it can break. A child matters partly because he is vulnerable. A life matters partly because it can end.

A conscious mind matters because it can suffer, love, imagine, remember, and be lost.

The soul was invented to make the self permanent. Science shows something different. The self is not permanent.

But it is precious.

What Sleep Teaches About Death

Sleep is sometimes called the little death.

That phrase is not exact, but it captures something important. In deep sleep, the conscious self is absent from experience. The world disappears, yet the body lives. The brain continues its work. Then consciousness returns.

Death is different.

In sleep, the brain remains alive and active. In death, the brain permanently loses the conditions needed to support consciousness. The process does not merely pause. It ends.

This is why sleep does not prove an afterlife. It teaches the opposite lesson: consciousness depends on the living brain. When the brain can return to waking function, the self returns. When the brain permanently cannot, there is no good evidence that the self continues elsewhere.

Religion tries to turn death into sleep with a promised awakening. Science gives no such guarantee.

This may feel painful, but the absence of eternal life does not make life meaningless. It makes life urgent. It asks us to take seriously the only consciousness we know we have. It asks us to care for the living. It asks us to protect the brain, the body, the child, the patient, the elderly, the injured, the grieving, and the vulnerable.

It asks us not to postpone love. It asks us not to wait for heaven to repair what we can repair here.

If consciousness can disappear every night, if dreams can create worlds, if anesthesia can silence awareness, if coma can hide or destroy the self, then we should not treat consciousness casually.

It is not a guaranteed possession. It is a fragile gift of the living brain.

The Comfort and Danger of the Soul

The soul gives comfort because it promises that the self cannot be truly lost. Sleep may come, but the soul remains. Disease may come, but the soul remains. Death may come, but the soul remains.

This is comforting.

But it can also be dangerous.

If people believe the soul is more important than the living brain, they may undervalue the body. If they believe heaven matters more than earthly suffering, they may neglect the urgent needs of this life. If they believe the true person is safe in a soul, they may misunderstand dementia, psychiatric illness, addiction, brain injury, trauma, or delirium.

A brain-based view teaches a more practical compassion.

Protect sleep. Prevent stroke. Treat addiction. Treat depression. Prevent head injury. Care for dementia. Research anesthesia. Improve ICU care. Rehabilitate brain injury. Support families. Reduce violence. Care for children.

Because the mind depends on the brain, these things are not secondary. They are central to human dignity.

A belief in the soul may comfort us after damage is done. A scientific understanding of the brain may help prevent damage, treat suffering, and protect life before it is too late.

That is not a small difference.

Awareness Without Magic

The altered states of consciousness discussed in this chapter all point toward one conclusion. Sleep changes awareness. Dreams create private worlds. Lucid dreams show that self-awareness can appear within sleep. Anesthesia can silence consciousness through brain effects. Coma and disorders of consciousness follow brain injury and brain dysfunction. Delirium, seizures, oxygen failure, and dementia can alter the conscious world.

These facts do not prove that every detail of consciousness is solved. They do not answer every philosophical question. They do not remove the mystery of subjective experience.

But they do remove the need for a soul as an explanation.

The soul adds mystery without explaining the evidence. The brain explains why awareness changes when the brain changes.

That is the stronger explanation.

A supernatural answer may feel comforting, but a natural answer fits reality.

And reality is where we must live.

Chapter Summary

This chapter examined altered states of consciousness: sleep, dreams, lucid dreaming, anesthesia, coma, and disorders of consciousness.

Sleep shows that ordinary awareness fades every night as the brain changes state. Dreams show that the sleeping brain can create vivid inner worlds from memory, emotion, and imagination. Lucid dreaming shows that self-awareness itself can vary during sleep. Anesthesia shows that consciousness can be profoundly altered or silenced by drugs that act on the nervous system. Coma and disorders of consciousness show that awareness depends on brain systems that can be damaged, hidden, or partially recover.

These states are difficult to explain if consciousness comes from an independent soul. They make far more sense if consciousness is a living process of the brain.

The fragility of awareness does not make human life meaningless. It makes consciousness more precious. It teaches us to protect the brain, care for the vulnerable, and value this life while we have it.

Key Takeaways

Sleep is a daily reminder that consciousness is not constant.

Dreams are vivid experiences created by the sleeping brain, not evidence that the soul travels.

Anesthesia shows that consciousness can be deeply altered by changing brain function.

Coma and disorders of consciousness reveal the dependence of awareness on injured or impaired brain systems.

The soul does not explain why awareness changes with sleep, drugs, injury, oxygen, seizures, or brain disease.

A brain-based view of consciousness increases compassion because it helps us protect, treat, and understand fragile human minds.

Transition to the Next Chapter

If consciousness depends on the brain, then another religious idea must be examined carefully: free will.

Religion often imagines a soul freely choosing good or evil, deserving reward or punishment from God. But human choice does not arise from nowhere. It is shaped by brain function, genes, childhood, habits, emotions, trauma, culture, education, and circumstance.

Does a brain-based view destroy responsibility?

Or can it make responsibility more humane?

The next chapter will examine free will without a ghost inside us.

Sources

Y. Nir and G. Tononi, "Dreaming and the Brain: From Phenomenology to Neurophysiology," *Trends in Cognitive Sciences*, 2010.

B. Baird et al., "The Cognitive Neuroscience of Lucid Dreaming," *Neuroscience & Biobehavioral Reviews*, 2019.

S. Scarpelli et al., "The Functional Role of Dreaming in Emotional Processes," *Frontiers in Psychology*, 2019.

A. N. Goldstein and M. P. Walker, "The Role of Sleep in Emotional Brain Function," *Annual Review of Clinical Psychology*, 2014.

J. Montupil et al., "The Nature of Consciousness in Anaesthesia," *British Journal of Anaesthesia*, 2023.

A. G. Hudetz, "Disconnecting Consciousness: Is There a Common Anesthetic End Point?" *Anesthesia & Analgesia*, 2016.

A. I. Luppi et al., "General Anaesthesia Decreases the Uniqueness of Brain Activity Across Individuals and Across Species," *Nature Human Behaviour*, 2025.

Q. He et al., "Advances in Understanding and Treating Disorders of Consciousness After Brainstem Injury," 2025.

Centers for Disease Control and Prevention, "About Sleep," 2024.

Chapter 6: Free Will Without a Ghost

Choice, Responsibility, and Compassion in a Brain-Based World

You choose.

You decide what to eat, what to say, where to go, whom to trust, when to speak, when to remain silent, when to forgive, when to resist, and when to walk away. You feel the weight of choice when you stand at a moral crossroad: tell the truth or hide it, help or ignore, control anger or release it, keep a promise or break it, act with courage or surrender to fear.

In those moments, it feels as if there is a private commander inside you, a little ruler sitting somewhere behind the eyes, examining options and making decisions freely.

Religion gave this commander a name.

The soul.

According to many religious traditions, the soul chooses good or evil. The soul obeys or disobeys. The soul sins or resists sin. The soul deserves reward or punishment. The soul is judged by God.

This picture is emotionally powerful because it makes human responsibility seem simple. A person does wrong because the soul chose evil. A person does good because the soul chose righteousness. A person is saved, damned, rewarded, or punished because of the choices of that inner spiritual self.

But real human life is not that simple.

You choose, but you do not choose from nowhere. Every decision carries your genes, childhood, brain development, habits, education, fears, desires, memories, sleep, pain, trauma, culture, social pressure, poverty or privilege, mental health, addiction, hormones, and the situation in front of you.

A hungry person does not choose in exactly the same way as a well-fed person. A frightened child does not choose in exactly the same way as a secure child. A person raised in violence does not choose in exactly the same way as a person raised in safety. A person with frontal brain injury, addiction, psychosis, dementia, severe depression, or extreme trauma may not choose with the same capacity as a healthy, rested, supported adult.

Choice is real, but choice is not magic.

This chapter asks a difficult question: if the mind is created by the brain, what happens to free will? Does a brain-based view make responsibility impossible? Does it mean criminals should not be accountable? Does it mean kindness and cruelty are just chemistry? Does it mean morality collapses?

No.

A brain-based view does not destroy responsibility. It makes responsibility more realistic, more compassionate, and more useful.

Understanding causes does not erase responsibility. It makes responsibility more intelligent.

It removes the ghost, but it does not remove the human being.

The Religious Idea of the Choosing Soul

Religion often imagines human choice as a struggle inside the soul. God commands. The devil tempts. The soul chooses. Then judgment follows.

This view is found in different forms across many religious traditions. Human beings are treated as morally responsible because they are believed to possess an inner spiritual agent capable of choosing obedience or disobedience.

This picture has a certain simplicity. If a person lies, the soul chose dishonesty. If a person steals, the soul chose greed. If a person commits violence, the soul chose evil. If a person doubts religion, the soul chose rebellion. If a person disobeys religious law, the soul chose sin.

This view gives religion enormous moral power because it turns behavior into spiritual evidence. Every action becomes a sign of the soul's condition. Obedience becomes virtue. Doubt becomes danger. Desire becomes temptation. Suffering may become punishment. Nonbelief may become guilt.

But this view has a serious flaw: it often ignores causes.

It may ignore brain disease, trauma, poverty, addiction, mental illness, childhood abuse, social pressure, education, fear, and biology. It may ignore the fact that people are not born as isolated souls floating in empty space. They are born as vulnerable bodies with developing brains, inside families, cultures, economies, histories, and power structures they did not choose.

A child does not choose the parents, country, caste, religion, language, poverty, or social world into which he is born. A young girl does not choose to be raised with shame. A boy does not choose to be beaten until anger becomes his first language. A child does not choose malnutrition, neglect, war, fear, or indoctrination. Yet all these things can shape the adult.

A moral system that ignores causes may feel clear, but it becomes cruel. It looks only at the final act and says, "The soul chose."

Science asks a deeper question: what created the conditions in which this choice became likely?

That question does not erase responsibility.

It improves it.

Why Choices Come From Causes

Every choice has a history.

You may choose coffee over tea in the morning. That choice may feel simple and free. But it carries years of habit, taste, culture, memory, sleep quality, stomach comfort, advertising, family patterns, and the brain's desire for alertness.

You may choose to speak gently instead of harshly. That choice may carry your temperament, moral training, love for the other person, fear of regret, past mistakes, emotional regulation, and the condition of your brain at that moment.

You may choose to read this book rather than another. That choice may carry curiosity, doubt, education, frustration with religion, respect for science, personal experience, and the larger story of your life.

Choices do not come from nothing. They arise from causes.

This does not make them unreal. A tree grows from soil, water, sunlight, roots, genes, weather, and time. The tree is not unreal because it has causes. A song comes from rhythm, breath, memory, instruments, training, culture, and emotion. The song is not fake because it has causes.

A choice can be real and still be caused.

This is where many people become confused. They think that if a choice has causes, it is not truly a choice. But that is not how ordinary life works.

A wise choice usually has good causes. A person chooses not to drive drunk because he understands risk, cares about others, remembers past harm, respects law, and has enough self-control to call for a ride. A person chooses to apologize because empathy, memory, love, and moral growth have shaped him. A person chooses to study because hope, ambition, discipline, fear of failure, and encouragement have influenced her.

Would we prefer choices without causes?

A choice with no cause would not be freedom. It would be randomness. If your decisions had no roots in your character, values, memory, reasoning, and concern for others, they would not be more free. They would be more chaotic.

The real question is not whether choices have causes. They do.

The question is: what kinds of causes shape better choices?

Good education, safe childhoods, emotional regulation, stable families, mental health care, reduced poverty, fair laws, honest information, freedom from religious fear, practice in reasoning, compassionate communities, protection from abuse, treatment for addiction, and real opportunity all matter.

A society that understands causes can reduce harm. A society that only blames souls may punish endlessly without preventing anything.

The Brain, Habit, Emotion, and Impulse

The brain does not make decisions as a single simple machine. Different systems contribute to choice. Emotion matters. Memory matters. Habit matters. Reward matters. Fear matters. Impulse control matters. Attention matters. Social pressure matters. Language matters. The body matters.

The brain is constantly weighing signals: danger, desire, pain, reward, habit, rules, memory, future consequences, and social meaning. Sometimes slower reasoning wins. Sometimes impulse wins. Sometimes fear wins. Sometimes compassion wins. Sometimes habit acts before reflection catches up.

This is why self-control is not simply a command issued by the soul. It is a capacity supported by brain development, emotional regulation, learning, sleep, stress level, and environment.

Anyone who has cared for children knows this. A two-year-old does not have the same self-control as a fifteen-year-old. A teenager does not have the same judgment as a mature adult. A tired child has less control than a rested child. A frightened child has less control than a secure child.

This is not because children have smaller souls. It is because their brains and self-regulation capacities are still developing.

The same principle applies throughout life. A person under extreme stress may make poorer decisions. A person with addiction may struggle against powerful reward circuits and learned craving. A person with severe depression may see no future. A person with mania may become impulsive and grandiose. A person with psychosis may act under beliefs that do not match reality. A person with dementia may lose judgment. A person with frontal lobe injury may lose inhibition. A person with sleep deprivation may become emotionally unstable.

None of this means that behavior has no consequences. It means that behavior must be understood in context.

The soul-based view often asks, "Why did this person choose evil?" A brain-based view asks, "What capacities were present, what capacities were impaired, what causes shaped this action, and how can future harm be prevented?"

That second question is morally better. It does not merely punish. It tries to understand. And understanding is the beginning of prevention.

What Neuroscience Can and Cannot Prove

In discussions of free will, one experiment is often mentioned: the Libet experiment.

In the early 1980s, Benjamin Libet and colleagues studied voluntary movement. Participants were asked to make a simple movement, such as flexing a wrist, while noting when they became aware of the urge or intention to act. Brain activity known as the readiness potential appeared before participants reported conscious awareness of deciding.

Some people took this to mean that free will is an illusion.

But that conclusion is too strong.

These experiments involved simple, artificial movements, not serious moral choices. Flexing a wrist in a laboratory is not the same as deciding whether to betray a friend, forgive an enemy, steal money, risk life for a child, or leave a harmful religion.

The timing of conscious intention is also difficult to measure. The meaning of the readiness potential remains debated. It may show preparation for action rather than a completed unconscious decision.

So we should be careful. This book does not claim that neuroscience has completely solved free will. It has not.

The better argument is more modest, more accurate, and more persuasive. Neuroscience shows that choices are brain-based. It shows that conscious decisions are part of a larger network of unconscious and preconscious processes. It shows that intention, impulse, attention, planning, habit, inhibition, and action depend on brain systems. It shows that the feeling of being a free chooser is not produced by a supernatural soul floating outside biology.

But neuroscience has not settled every philosophical debate about free will.

That is fine.

We do not need to solve every philosophical debate to reject the religious picture of a ghost-like soul choosing outside nature. The evidence already shows that human choice is embodied, conditioned, brain-based, and shaped by causes.

That is enough to weaken the soul-based view.

Freedom Is Not Magic

What do we mean by freedom?

If freedom means a soul making choices completely outside biology, history, personality, emotion, brain function, and circumstance, then there is no good evidence for that kind of freedom.

But that is not the only meaning of freedom.

A person can be meaningfully free when he can think, reflect, understand consequences, control impulses, act according to values, learn from mistakes, and respond to reasons. A person is less free when these capacities are damaged.

A prisoner is less free because external force restricts action. A person with severe addiction may be less free because craving narrows choice. A person with psychosis may be less free because reality testing is impaired. A person with dementia may be less free because memory and judgment are damaged. A person raised under terror may be less free because fear shapes the mind. A person indoctrinated from childhood may be less free because certain thoughts were made dangerous before reason could examine them.

Freedom is not the absence of causes. Freedom is the presence of certain capacities: the capacity to understand, reflect, imagine alternatives, control impulses, learn, act without coercion, revise belief, care about consequences, and respond to evidence.

These capacities depend on the brain, education, society, and health.

This view is more useful than the religious idea of a mysterious soul. It allows us to ask how freedom can be increased.

Teach children to think. Protect them from fear. Improve education. Reduce abuse. Treat mental illness. Treat addiction. Encourage honest doubt. Protect free speech. Build fair laws. Give people real opportunities. Help people develop self-control.

That is how freedom grows in the real world: not by declaring that every soul is equally free while ignoring the chains placed on bodies and minds.

Responsibility Without Revenge

Some people fear that if behavior has causes, no one can be held responsible.

This fear is understandable, but it rests on a mistake.

Responsibility does not require belief in an uncaused soul. Responsibility is part of how human beings live together.

If a person harms another, society must respond. Victims must be protected. Dangerous behavior must be stopped. People must learn that actions have consequences. Restitution may be needed. Treatment may be needed. Rehabilitation may be possible. In some cases, separation from society may be necessary.

But responsibility should not be confused with revenge.

A soul-based view often treats wrongdoing as a stain deserving punishment. A brain-based view asks what response will reduce future harm, protect the vulnerable, and encourage better behavior.

This is not softness. It is seriousness.

Revenge looks backward and says, "You deserve suffering." Justice looks backward and forward. It says, "Harm was done. The victim matters. The truth matters. The community matters. The causes matter. The future matters. What response is necessary to protect life and reduce further harm?"

A person is not automatically innocent because the brain was involved. The brain is always involved.

The real questions are more specific. Could the person understand what he was doing? Could the person understand that it was wrong? Could the person control behavior? Was there severe impairment? Is the person dangerous? Can treatment reduce risk? What protects the victim and society? What is fair?

A brain-based view does not abolish these questions. It makes them more precise.

It moves us away from supernatural blame and toward intelligent accountability.

Justice With Prevention and Compassion

If we want a better society, we should care not only about punishment after harm, but prevention before harm.

This is where a brain-based view becomes morally powerful. If childhood trauma increases later risk, reduce childhood trauma. If lead exposure harms brain development and behavior, remove lead from homes and water. If untreated addiction increases crime and suffering, treat addiction. If severe mental illness leads to crisis, improve care before crisis. If poverty and neglect damage opportunity, reduce poverty and neglect. If poor education leaves people with fewer tools for self-control and reasoning, improve education. If religious indoctrination teaches fear, shame, hatred, or obedience to harmful authority, challenge it. If social media feeds rage, tribalism, and misinformation, teach critical thinking.

A soul-based view may wait until after harm and then condemn the sinner. A scientific moral view asks how to reduce harm before it happens.

That is not an excuse for wrongdoing. It is a more mature form of responsibility.

Imagine two societies.

One says, “People do evil because their souls are wicked. Punish them.”

The other says, “People are responsible for harm, but harmful behavior has causes. Let us protect victims, hold offenders accountable, and also reduce the conditions that produce violence, addiction, abuse, and cruelty.”

Which society is more moral?

The second is.

Because the second cares about actual suffering, not just moral drama. It wants fewer victims, fewer damaged children, fewer addicts, fewer assaults, fewer murders, fewer broken families, fewer prisons, fewer ruined lives.

That is morality in the real world.

Sin, Shame, and the Burden of Religious Blame

Religion often uses the language of sin to describe human weakness. Desire becomes sin. Doubt becomes sin. Anger becomes sin. Sexuality becomes sin. Disobedience becomes sin. Nonbelief becomes sin.

This language can create deep shame, especially in children. A child may learn to fear thoughts before understanding them. A teenager may learn to hate his own body. A doubter may believe that honest questioning is rebellion against God. A person with depression may feel spiritually weak. A person with addiction may feel morally filthy rather than medically and psychologically trapped.

A brain-based view offers relief from unnecessary shame. It says: you are a human being with a brain and body. You have impulses because you are biological. You have fears because you are vulnerable. You have desires because you are alive. You have habits because the brain learns. You have intrusive thoughts because minds generate thoughts. You may suffer depression, anxiety, addiction, trauma, obsession, or confusion because brains and bodies can suffer.

This does not mean every impulse should be followed. It does not mean every desire is wise. It does not mean every behavior is acceptable.

But it removes the poison of supernatural shame.

Instead of saying, "Your soul is stained," it asks, "What is happening, what harm is involved, what help is needed, and what better choice can be built?"

That is more humane.

A morality based on understanding can still be firm. It can say no. It can set limits. It can protect others. It can demand repair.

But it does not need to crush a person under metaphysical guilt.

A person can take responsibility without believing he is spiritually contaminated.

Why Understanding Causes Makes Morality Deeper

Some people think that explaining behavior makes morality weaker. They say, "If you explain cruelty, you excuse it."

That is not true.

To explain is not to excuse. A doctor may explain how smoking damages the lungs. That does not mean smoking is good. A mechanic may explain why brakes failed. That does not mean the crash did not matter. A historian may explain how propaganda led to violence. That does not mean violence was justified. A neuroscientist may explain how impulse control fails. That does not mean harm should be ignored.

Understanding causes helps us respond intelligently.

If a child steals food because he is hungry, the response should differ from a wealthy adult stealing for greed. If a person acts under psychosis, the response should differ from a person acting with clear planning. If a teenager makes a reckless decision, the response should consider development. If an addict commits harm to obtain drugs, accountability may be needed, but treatment is also essential. If a violent ideology trains people to hate, we must address the ideology, not only punish the final act.

Moral judgment becomes deeper when it includes causes.

Religious morality often prefers simplicity: good soul, bad soul, obedience, sin, punishment, reward. Human reality is more complicated.

A serious morality must be able to hold several truths at once. The victim matters. The offender has causes. The harm was real. The future must be protected. The person may still be capable of change. Some people must be restrained. Some people can be rehabilitated. Prevention is better than revenge. Compassion and accountability can exist together.

This is not moral weakness.

It is moral maturity.

The Freedom to Change

A brain-based view does not make change impossible. It makes change understandable.

The brain is not fixed like stone. It changes with learning, practice, experience, attention, therapy, habit, relationship, and environment.

A person can learn to pause before anger. A person can recover from addiction. A person can leave religious fear. A person can build new habits. A person can become more compassionate. A person can question inherited prejudice. A person can become less reactive. A person can learn emotional regulation. A person can repair harm. A person can grow.

This is one of the most hopeful truths about human beings.

We are shaped by causes, but we can also become part of the causes that shape our future. A person who understands his anger can practice restraint. A person who understands addiction can seek treatment. A person who understands indoctrination can begin to question. A person who understands anxiety can learn skills. A person who understands prejudice can challenge it. A person who understands mortality can choose to live more honestly.

This is not supernatural freedom. It is natural freedom.

It is the freedom of a brain capable of reflection, learning, self-correction, and growth.

Religion sometimes tells people that transformation comes from surrender to God. Science and human experience show another path. Transformation can come from knowledge, practice, therapy, love, social support, moral courage, honest self-examination, and changed conditions.

No ghost is needed.

No divine intervention is required.

Free Will Without God

Many religious people believe that without God, free will has no foundation. But this is backwards. The religious idea often creates more problems than it solves.

If God is all-knowing, then God already knows every future choice. If God created the person knowing every choice that person would make, then how free is the person in any ultimate sense? If God creates people knowing some will be damned, what kind of justice is that? If God intervenes in human lives, answers prayers, hardens hearts, sends temptations, allows demons, or controls destiny, where exactly does human freedom begin and end?

Religion does not solve free will. It complicates it.

A natural view is clearer. Human beings are organisms with brains capable of perception, memory, emotion, reasoning, self-control, imagination, and social learning. Our choices arise from these capacities. They are shaped by causes. They are limited by conditions. They can improve with knowledge and practice. They can fail under disease, coercion, addiction, fear, and injury.

This view does not give us absolute freedom.

But absolute freedom was never real.

It gives us human freedom: limited, embodied, unequal, fragile, and expandable.

That is the kind of freedom that matters in real life. Not freedom from nature, but freedom within nature.

Chapter Summary

This chapter examined free will without the idea of a soul.

Religion often imagines a soul choosing good or evil outside biology and then deserving divine reward or punishment. But human choices do not come from nowhere. They arise from brain function, genes, childhood, habits, emotion, trauma, education, culture, social pressure, health, and circumstance.

Neuroscience has not simply disproved free will. Experiments about conscious intention raise important questions, but they do not settle every philosophical issue. The stronger and more careful conclusion is that human choice is brain-based, conditioned, and shaped by causes.

This does not destroy responsibility. It changes how we understand responsibility. A brain-based view supports accountability, protection of victims, prevention of harm, treatment where appropriate, rehabilitation when possible, and compassion without naivety.

Freedom is not magic. It is the capacity to understand, reflect, control impulses, respond to reasons, learn from consequences, and act without coercion. These capacities depend on the brain and the world around us.

A natural view of choice does not make morality weaker. It makes morality deeper, more humane, and more effective.

Key Takeaways

Human choices are real, but they do not arise from a supernatural soul outside biology.

Every decision has causes: brain function, memory, emotion, habit, childhood, culture, health, and circumstance.

Neuroscience has not simply disproved free will, but it strongly challenges the religious idea of a ghost-like chooser inside the body.

Responsibility does not require belief in an uncaused soul.

A brain-based view allows justice to focus on protection, prevention, accountability, treatment, and rehabilitation rather than revenge.

Understanding causes does not excuse harm. It helps reduce future harm.

Freedom is not the absence of causes. It is the presence of capacities such as reflection, self-control, learning, and response to reasons.

Transition to the Next Chapter

If choice does not come from a supernatural soul, then the next question becomes unavoidable.

Where does morality come from?

Religion says morality comes from God. It says commandments, scriptures, heaven, hell, sin, and divine judgment are necessary for human goodness.

But is that true?

Long before theology, living beings cared for offspring, formed bonds, cooperated, warned one another, shared, protected, and reacted to harm. Human morality did not appear suddenly from the sky. It grew from life together.

The next chapter will examine morality before God.

It will show how care, cooperation, empathy, fairness, and reason made ethics possible without divine command.

Sources

Daniel Dennett, *Freedom Evolves.*

Michael Gazzaniga, *Who's in Charge? Free Will and the Science of the Brain.*

Robert Sapolsky, *Behave: The Biology of Humans at Our Best and Worst.*

Robert Sapolsky, *Determined: A Science of Life Without Free Will.*

Sam Harris, *Free Will.*

M. Brass, P. Haggard, and others, "Why Neuroscience Does Not Disprove Free Will," *Neuroscience & Biobehavioral Reviews*, 2019.

C. D. Frith and P. Haggard, "Volition and the Brain: Revisiting a Classic Experimental Study," *Trends in Neurosciences*, 2018.

A. Schurger et al., "What Is the Readiness Potential?" *Trends in Cognitive Sciences*, 2021.

J. Shepherd, "Scientific Challenges to Free Will and Moral Responsibility," *Philosophy Compass*, 2015.

A. Bigenwald and F. Chambon, "Criminal Responsibility and Neuroscience: No Revolution Yet," *Frontiers in Psychology*, 2019.

Stanford Encyclopedia of Philosophy, "Free Will."

Chapter 7: Morality Before God

How Care, Cooperation, and Fairness Came Before Commandments

A mother does not need scripture before she protects her child. A child does not need theology before he feels that something is unfair. A friend does not need a priest before feeling the pain of betrayal. A stranger does not need a doctrine of heaven before pulling another person from danger.

Long before commandments were written, living beings cared for their young, formed bonds, warned one another of danger, shared food, defended groups, punished cheaters, and grieved losses.

Morality did not begin in a temple. It began in life together.

Religion often claims that morality came down from heaven, fully formed, like a set of divine instructions handed to humanity by God. According to this view, without God there is no good and evil. Without commandments there is no moral law. Without heaven and hell there is no reason to be kind. Without divine judgment, human beings would collapse into selfishness, cruelty, and chaos.

This claim is repeated so often that many people accept it without asking whether it is true.

But look honestly at human life. Parents protect children before children understand religion. People feel guilt before they study theology. People feel gratitude before they read scripture. People feel outrage at cruelty before they can explain moral philosophy. People comfort the grieving without first solving the question of God.

People know that torture hurts because pain is real, not because a holy book says so. They know that betrayal wounds trust because relationships are real, not because a priest defines sin. They know that hunger, humiliation, violence, fear, loneliness, and injustice matter because conscious beings suffer.

Morality begins there: in suffering, care, dependence, and the fact that we live with other beings whose lives can be helped or harmed.

This does not mean morality is simple. It does not mean human beings are naturally perfect. We are not. We are capable of kindness and cruelty, loyalty and betrayal, courage and cowardice, compassion and tribal hatred.

But the roots of morality are not supernatural. They are human, social, biological, emotional, and rational.

Religion did not create morality. Religion inherited moral instincts, shaped them, claimed them, and sometimes distorted them.

Goodness is older than God.

Care Came Before Commandments

The first moral fact of human life is dependence.

A newborn cannot survive alone. A baby cannot feed itself, protect itself, warm itself, carry itself, or understand danger. A child begins life completely dependent on care.

This simple fact shaped human morality more deeply than any scripture. Before commandments, there was the cry of a child. Before theology, there was the body of the parent responding. Before doctrines of sin and salvation, there was feeding, holding, protecting, teaching, and comforting.

Care is not an abstract idea. It is one of the oldest foundations of moral life.

A mother who wakes at night to feed a crying child is not calculating eternal reward. A father who runs toward danger to protect his child is not first solving a religious argument. A grandmother who comforts a frightened child is not performing metaphysics. She is responding to vulnerability.

This does not require God. It requires attachment, empathy, memory, body, and love.

In mammals, care for offspring is biologically deep. In humans, it becomes even richer because childhood is long. Human children require years of protection, teaching, language, emotional support, and social learning. A species with helpless children cannot survive without care.

This is one reason morality could not have begun as a set of abstract rules. It began as practical life: feed the child, protect the child, do not abandon the child, do not let the group collapse, do not let violence destroy trust, do not let selfishness ruin cooperation.

The earliest moral world was not written on stone tablets. It was written in hunger, danger, injury, birth, nursing, fear, play, and family bonds.

This is why the religious claim that morality requires commandments is so shallow. Commandments may express moral ideas. They may enforce moral ideas. They may also distort moral ideas. But they did not create the need for moral behavior.

Life created that need.

A child in pain does not need a commandment to make the pain morally important.

The pain itself is enough.

Cooperation Came Before Theology

Human beings are not solitary creatures. We survive together.

A lone human is weak compared with many animals. We do not have the claws of a tiger, the speed of a deer, the strength of a gorilla, or the armor of a turtle. Our great strength is cooperation.

We hunt together, gather together, build together, defend together, raise children together, share knowledge together, remember together, and teach together. A human group that could cooperate had a better chance of survival than a group destroyed by constant suspicion, selfishness, and violence.

This is one of the deep roots of morality.

Morality helps groups live together. Do not steal from those who trust you. Do not betray those who depend on you. Do not abandon the helpless. Do not lie when truth is necessary. Do not murder members of your own group. Share when sharing protects the group. Punish those who destroy trust. Honor those who sacrifice. Teach children what behavior is acceptable.

These rules did not need to fall from the sky. They emerged because social life required them.

This does not mean morality is only a survival strategy. Human morality later becomes richer: justice, rights, dignity, equality, compassion for strangers, concern for animals, and care for future generations. But cooperation is one of its deepest foundations.

Even today, most moral anger is connected to damaged cooperation. We are angry when someone cheats, lies, takes without giving, abuses power, harms the innocent, or receives unfair advantage. These reactions are not random. They protect the conditions of shared life.

Religion often says, "God gave morality."

But the more realistic answer is this: life together required morality.

Groups that could not develop rules of trust, restraint, cooperation, care, and fairness could not flourish.

God was not needed to create this need.

Fairness Came Before Scripture

Fairness is one of the first moral feelings children show.

Anyone who has watched children knows this. A child may not understand law, theology, or philosophy, but the child understands, "That is not fair."

One child receives a larger piece of cake. Another gets a toy first. One is punished while another escapes punishment. One is praised for the same thing another was scolded for. Immediately, the protest comes: "That is not fair!"

Fairness is not a complete moral system, but it is a powerful moral beginning.

We should not exaggerate. Babies do not possess adult morality. They do not have a full theory of justice, rights, law, consent, equality, or human dignity. Children must be taught, guided, corrected, and helped to think beyond themselves.

But the early signs matter.

Children are not blank moral stones waiting for priests to write commandments on them. They come into the world with social capacities, emotional responses, attachment needs, and developing expectations about care, help, harm, and fairness.

Fairness also appears in nonhuman animals in simpler forms. This does not mean other animals have human morality. They do not write laws, debate rights, or build ethical philosophies. But some building blocks of moral life are older than religion and older than humanity itself.

Religion may claim that fairness comes from God. But fairness appears to grow from social living.

If two individuals cooperate, unfairness threatens cooperation. If one always takes more, trust breaks. If one cheats, resentment grows. If power becomes abuse, the group becomes unstable.

Fairness is not merely a divine idea. It is a social necessity.

This is why fairness can be felt before it is explained. A child may not understand moral philosophy, but he knows when he has been treated unfairly. An adult may not know the language of ethics, but she knows when dignity has been violated. A worker may not quote a scripture or a philosopher, but he knows when his labor has been exploited. A woman may not use the vocabulary of rights, but she knows when her freedom has been denied.

Fairness begins in lived experience.

Not in heaven.

Moral Emotions Came Before Priests

Morality is not only reason. It is also emotion.

Imagine a person who harms a friend and feels no guilt. Imagine a person who receives kindness and feels no gratitude. Imagine a person who sees a child suffering and feels no compassion. Imagine a person who humiliates the weak and feels no shame.

Such a person may understand moral rules intellectually, but something important is missing.

Moral emotions are part of what make morality human. Guilt tells us we may have harmed someone or violated our own standards. Gratitude tells us we have received care, help, or generosity. Shame, though dangerous when excessive, can warn us that we have fallen below social or personal expectations. Compassion moves us toward another person's suffering. Anger at injustice pushes us to resist harm. Love binds us to the well-being of others.

These emotions did not require priests to invent them. They are part of social life.

Of course, moral emotions can mislead. Shame can become toxic. Guilt can be manipulated. Disgust can be attached to harmless differences. Religious systems have often exploited guilt and shame, especially around sexuality, doubt, caste, gender, and obedience.

Moral emotions are powerful, but they require guidance by reason and compassion.

Still, without moral emotion, morality becomes dry and weak. A person may know that hunger is bad, but compassion makes hunger urgent. A person may know that injustice is wrong, but moral anger makes injustice intolerable. A person may know that betrayal is harmful, but guilt may prevent him from repeating it.

Religion often claims these emotions as evidence of God inside the soul. But they are better understood as part of human social evolution, brain development, family life, and culture.

We feel moral emotions because we are not isolated souls. We are social beings. We depend on others. We can be harmed by others. We can harm others. We need trust. We need belonging. We need care.

Moral emotions help regulate that shared life. They are not perfect, but they are real.

Helping, Sharing, and Comforting Came Before Doctrine

Human morality is visible not only in grand sacrifices, but also in small acts.

A child picks up an object an adult has dropped. A person shares food. A friend comforts someone who is crying. A neighbor checks on an elderly person. A stranger holds a door. A passerby calls for help after an accident.

These small acts are not trivial. They are the everyday fabric of moral life.

Helping responds to need. Sharing responds to lack. Comforting responds to distress. These are not religious inventions. They are social responses to vulnerability.

A child does not need to understand God before handing a dropped spoon back to an adult. A friend does not need theology before comforting grief. A neighbor does not need scripture before bringing food to a sick family.

Religion may encourage such acts, and when it does, that is good. But it did not create the human capacity for them.

The capacity comes from our nature as social, emotional, learning beings. The direction and expansion of that capacity come from culture, education, reflection, and moral growth.

That is why morality can improve.

If morality were simply a fixed divine code, moral progress would be hard to explain. But if morality is a human capacity shaped by experience, reason, and expanding concern, then moral progress makes sense.

We can learn to care beyond family, beyond tribe, beyond caste, beyond religion, beyond nation, and even beyond species.

This is not morality falling from heaven.

It is morality growing through human struggle.

Reason Enlarges the Moral Circle

Care often begins close to home: family first, children first, friends first, tribe first, village first, nation first, religion first.

This is natural, but it is also dangerous.

The same human mind that loves its own group may hate another group. The same loyalty that protects family may justify cruelty toward strangers. The same religion that teaches kindness within the group may teach suspicion or violence toward outsiders.

This is why morality cannot depend only on instinct or tradition. It needs reason.

Reason asks uncomfortable questions. Why should my child matter more than another child simply because he is mine? Why should a person born into one caste have less dignity than another? Why should women have fewer rights than men? Why should a person's religion determine whether I treat him with compassion? Why should sexual minorities be punished for harmless private life? Why should animals suffer unnecessarily for human pleasure? Why should future generations pay for our selfishness today?

These questions expand the moral circle.

The moral circle is the range of beings we believe deserve moral concern. In ancient life, this circle was often narrow: family, tribe, religion, caste, nation. Outsiders could be ignored, enslaved, conquered, or killed.

Moral progress often means widening that circle from family to stranger, from tribe to humanity, from men to women, from upper caste to all castes, from believer to unbeliever, from ruler to subject, from adult to child, and from human beings to animals capable of suffering.

Religion has sometimes helped this expansion, but often it has resisted it. Sacred texts and religious authorities have defended slavery, caste, patriarchy, persecution, and inequality. Moral progress frequently required people to question inherited religious authority.

Reason does not make us perfect. Human beings can use reason selfishly. They can rationalize cruelty. But reason gives us a tool for self-correction.

It allows us to ask whether a tradition causes harm. It allows us to compare suffering, examine evidence, notice hypocrisy, and say, "If I would not accept this treatment for myself, why should I impose it on another?"

This is one of the most important foundations of secular morality: not blind instinct, not blind obedience, but care corrected by reason.

Religion Does Not Own Goodness

Religious people often do good things. They feed the hungry, care for the sick, comfort the grieving, build schools, give charity, visit the lonely, help neighbors, and risk their lives for others.

This should be acknowledged honestly.

The argument of this book is not that religious people cannot be moral. That would be false, unfair, and foolish.

The argument is that religion does not own morality.

A religious person who feeds a hungry child is doing something good because hunger is real and feeding reduces suffering. A religious person who comforts a grieving widow is doing something good because grief is real and comfort matters. A religious person who protects the weak is doing something good because harm is real and protection matters.

The goodness of these acts does not depend on God. It depends on the human reality involved.

If an atheist feeds the same hungry child, the child is fed. If an atheist comforts the same grieving widow, comfort is given. If an atheist protects the same weak person, harm is prevented.

The moral value is in the act and its human meaning, not in the religious label attached to it.

This is devastating to the claim that morality requires God.

If kindness reduces suffering, it is good whether done by a believer or nonbeliever. If cruelty causes suffering, it is wrong whether done by a believer or nonbeliever.

Religion may motivate some people to do good. It may provide community, ritual, encouragement, and moral language. But motivation is not ownership.

A person may exercise because of a doctor's advice, a friend's encouragement, or personal discipline. That does not mean the doctor, friend, or discipline invented muscle.

Similarly, religion may sometimes encourage moral behavior, but it did not invent care, empathy, fairness, or cooperation. These were already part of human social life.

The Problem of Religious Tribal Morality

One of the dangers of religion is that it often narrows morality to the group.

Be kind to believers. Distrust unbelievers. Help those inside the faith. Convert those outside it. Marry within the community. Obey sacred authority. Defend the honor of the group. Punish those who leave. Condemn those who doubt.

This is not universal morality. It is tribal morality with divine decoration.

When morality is tied to religious identity, outsiders can be treated as threats. The unbeliever becomes dangerous. The apostate becomes traitor. The heretic becomes poison. The woman who disobeys becomes shame. The lower caste person becomes spiritually inferior. The sexual minority becomes sinful. The critic becomes enemy of God.

This is how religion can turn ordinary moral instincts upside down. Compassion becomes selective. Cruelty becomes duty. Doubt becomes sin. Obedience becomes virtue. Persecution becomes holy.

This does not happen in every religious person. Many believers rise above the tribalism of their traditions. But they do so by allowing human conscience to soften religious boundaries.

That again proves the point.

The best morality is not obedience to group identity. The best morality is concern for suffering, dignity, freedom, fairness, and truth, even when the person suffering does not belong to our group.

Science-based atheism has an advantage here. It does not need to divide humanity into saved and damned, believer and infidel, pure and impure, chosen and rejected.

It can begin with a simpler and broader truth: this is a conscious being; this being can suffer; that suffering matters.

No religion is needed for that.

Morality as Cooperation, Not Command

One modern way of understanding morality is to see it as a set of solutions to the problems of living together.

Who gets what? Who helps whom? Who can be trusted? Who deserves punishment? Who has authority? How should conflicts be settled? How should families share responsibilities? How should courage, generosity, loyalty, fairness, and honesty be valued?

This does not mean all morality is simple self-interest. It means that many moral rules solve recurring problems that arise when social beings must live together.

How do we live together without destroying one another? How do we protect trust? How do we reduce violence? How do we distribute resources? How do we care for children? How do we punish exploitation? How do we reward sacrifice? How do we keep promises?

These are not heavenly problems. They are human problems. And human beings created moral systems to answer them.

Some answers were good. Some were terrible. Some improved over time. Some became frozen by religion.

A secular morality has one great advantage: it can admit that old answers may be wrong.

A divine commandment is difficult to revise. A human moral rule can be examined, criticized, and improved.

That is not weakness.

That is progress.

Goodness Without Heaven or Hell

Many religious people ask, “Why be good without heaven or hell?”

The question itself reveals a problem.

If a person helps the poor only to gain heaven, is that the highest form of compassion? If a person avoids murder only to avoid hell, is that the deepest form of morality? If a person tells the truth only because God is watching, is that mature honesty?

Rewards and punishments can influence behavior. Society uses them. Parents use them. Schools use them. Laws use them. But the highest moral life cannot be built only on reward and punishment.

A good person should care that another being suffers. A good person should care that trust has been broken. A good person should care that dignity has been violated. A good person should care that injustice damages lives. A good person should care that cruelty leaves wounds.

The reason to be good is not that heaven pays well. The reason to be good is that suffering is real and our actions matter.

This is a more mature morality.

A doctor should not treat a patient well only because God is watching. A parent should not care for a child only because heaven rewards it. A citizen should not defend justice only because hell threatens. A person should do these things because human lives matter.

The moral center is not supernatural surveillance.

The moral center is the reality of harm and care.

Moral Progress Did Not Fall From Heaven

If morality came perfectly from God, one might expect religious societies to have always known the highest moral truths.

But history shows otherwise.

Human beings have slowly and painfully expanded moral concern. Slavery was defended for centuries. Caste hierarchy was defended as sacred order. Women were denied education, property, leadership, and independence. Children were beaten in the name of discipline. Heretics were burned. Apostates were killed. Blasphemers were punished. Sexual minorities were persecuted. Other religions were demonized.

Many of these injustices were defended by religion.

Moral progress often came not from obeying sacred tradition, but from challenging it.

People had to ask: Is this suffering justified? Is this hierarchy fair? Is this cruelty necessary? Is this person less human? Is this command moral simply because it is old? Is this scripture above compassion?

The more humanity asked such questions, the more morality improved.

This does not mean secular people were always good or religious people always bad. History is not that simple. Secular ideologies have also caused great harm when they became dogmatic, tribal, authoritarian, or indifferent to human suffering.

The lesson is not that atheists are automatically moral. The lesson is that morality improves when no idea is protected from criticism.

Religious morality often freezes ancient customs by calling them divine. Secular morality can revise itself because it admits human responsibility.

That is its strength.

The Foundation Beneath Morality

If morality does not come from God, where does it stand?

It stands on the reality of conscious life.

A being that can suffer matters. A being that can feel fear matters. A being that can be humiliated matters. A being that can love and grieve matters.

A child's pain matters before any scripture speaks. A woman's dignity matters before any priest approves. A slave's suffering matters even if a holy book permits slavery. An oppressed caste person's humanity matters even if tradition denies it. A nonbeliever's life matters even if religion condemns disbelief.

This is the foundation: not God, not heaven, not hell, not command, but the reality of conscious beings who can be harmed or helped.

From that foundation, morality grows. Care responds to vulnerability. Fairness responds to conflict. Justice responds to harm. Freedom responds to oppression. Honesty responds to trust. Compassion responds to suffering. Reason helps us correct error. Evidence helps us see consequences. Human dignity helps us resist cruelty.

This foundation is not perfect because humans are not perfect. But it is better than obedience to invisible authority.

It is open to correction. It is tied to real life. It does not require belief. It does not divide humanity into saved and damned.

It asks a simple question: what kind of world reduces suffering and protects dignity?

That is enough to begin morality.

Chapter Summary

This chapter examined the origins of morality without God.

Religion often claims that morality came from divine command, but care, cooperation, fairness, empathy, guilt, gratitude, compassion, and moral concern are rooted in human social life. Before commandments, there were vulnerable children, families, groups, danger, trust, betrayal, hunger, injury, and the need to live together.

Human morality grew from the problems and possibilities of social life. Cooperation made survival possible. Fairness helped protect trust. Moral emotions helped regulate behavior. Helping, sharing, and comforting emerged as responses to need, deprivation, and distress. Reason later expanded the moral circle beyond family, tribe, religion, caste, and nation.

Religion may motivate some moral behavior, but it does not own morality. Kindness is good because it reduces suffering. Cruelty is wrong because it causes suffering. These facts do not depend on God.

Morality did not begin in heaven. It began in life together.

Key Takeaways

Care came before commandments.

Cooperation came before theology.

Fairness came before scripture.

Moral emotions came before priests.

Human beings are social animals, and cooperation made morality necessary.

Religion may encourage morality, but it did not create care, empathy, fairness, or cooperation.

Religious morality can become tribal when it protects only the in-group and condemns outsiders.

A secular morality can begin with a simple truth: conscious beings can suffer, and suffering matters.

Transition to the Next Chapter

If morality existed before religion, then one of religion's most common accusations becomes weaker.

Religious believers often ask, "Without God, what stops people from doing anything they want?"

But that question misunderstands morality.

Fear of hell is not the same as goodness. Obedience to divine surveillance is not the same as moral maturity.

The next chapter will answer the accusation directly.

Without God does not mean without goodness.

In fact, morality based on human suffering, compassion, fairness, and responsibility may be deeper than morality based on fear.

Sources

Michael Tomasello, *A Natural History of Human Morality*.

Michael Tomasello, "Origins of Human Cooperation and Morality," *Annual Review of Psychology*, 2013.

Frans de Waal, *The Bonobo and the Atheist*.

Sarah F. Brosnan and Frans B. M. de Waal, "Evolution of Responses to (Un)Fairness," *Science*, 2014.

P. Limone and G. Toto, "Origin and Development of Moral Sense: A Systematic Review," 2022.

K. A. Dunfield, "A Construct Divided: Prosocial Behavior as Helping, Sharing, and Comforting," *Frontiers in Psychology*, 2014.

Oliver Scott Curry, Daniel Austin Mullins, and Harvey Whitehouse, "Is It Good to Cooperate? Testing the Theory of Morality-as-Cooperation in 60 Societies," *Current Anthropology*, 2019.

Luca Surian, "Core Moral Concepts and the Sense of Fairness in Human Infants," 2025.

Richard Joyce, *The Evolution of Morality*.

Chapter 8: The False Fear of Atheism

Why Without God Does Not Mean Without Goodness

If the only reason a person does not lie, steal, rape, or kill is fear of God, that is not deep morality. It is fear of punishment.

If the only reason a person helps the poor is hope of heaven, that is not pure compassion. It is a transaction.

If the only reason a person behaves decently is because he believes he is being watched by an invisible judge, that is not moral maturity. It is surveillance.

Religion often makes one of its most serious accusations against atheism: without God, anything is permitted; without God, there is no morality; without heaven and hell, human beings will do whatever they want; without divine judgment, kindness, honesty, sacrifice, and justice have no foundation.

This accusation is powerful because it frightens people. It makes atheism appear dangerous. It suggests that unbelief does not merely reject God, but opens the door to selfishness, cruelty, violence, and moral collapse.

For many believers, this fear is more powerful than any argument about evolution, the Big Bang, or the age of the earth. They may accept some science, but still ask, "Without God, why would anyone be good?"

This chapter answers that question directly.

People can be good without God because morality does not begin with God. Morality begins with conscious beings who can suffer. It begins with children who need protection. It begins with trust that can be broken. It begins with pain that can be relieved. It begins with cruelty that wounds. It begins with fairness, empathy, cooperation, gratitude, love, responsibility, and reason.

God is not required for any of this.

The fear that atheism destroys morality is one of religion's oldest and most successful claims.

It is also false.

The Old Accusation Against Atheists

The accusation against atheists has been repeated for centuries.

Atheists are immoral. Atheists cannot be trusted. Atheists have no reason to be good. Atheists secretly want a world without rules. Atheists reject God because they want to sin. Atheists will destroy family, society, and civilization.

These claims are not usually based on careful observation. They are often inherited prejudices. Many people are taught from childhood that belief in God is the foundation of goodness. If that is repeated often enough, disbelief begins to look like moral danger.

The accusation works because it attacks character, not just belief. It does not merely say, "You are mistaken about God." It says, "You are dangerous." It says, "You cannot be fully moral." It says, "You cannot be trusted with children, family, society, or power."

That is why this issue matters.

The question of atheism and morality is not just philosophical. It affects real people. Nonbelievers may be treated as suspicious, cold, selfish, arrogant, or corrupt even when they live kind and responsible lives. In some societies, open atheism can damage family relationships, social standing, career opportunities, safety, and even legal rights.

Religion often asks atheists to prove that they can be moral.

But perhaps the burden should be reversed.

If religious morality is truly superior, why has religion so often supported slavery, caste, patriarchy, persecution, holy war, blasphemy laws, forced conversion, fear of hell, and punishment of honest doubt? If belief in God guarantees morality, why are prisons not empty of believers? If religion makes people kind, why do religious groups so often hate other religious groups? If divine command is the foundation of goodness, why have so many divine commands defended cruelty?

The question is not whether religious people can be good. Of course they can.

The question is whether God is necessary for goodness.

The answer is no.

Fear Is Not the Same as Goodness

Fear can control behavior.

A child may obey because he fears punishment. A worker may follow rules because he fears losing a job. A driver may slow down because he sees a police car. A person may avoid crime because he fears prison.

Fear can restrain people. Sometimes restraint is necessary. Laws, consequences, and punishment have a role in society.

But fear is not the highest form of morality.

A person who does not steal only because he fears being caught is not deeply honest. A person who does not hurt others only because he fears hell is not deeply compassionate. A person who helps others only because he wants heaven is not truly selfless.

Religion often presents heaven and hell as moral tools. Be good and you will be rewarded. Be bad and you will be punished. God is watching. Nothing is hidden. Every thought, desire, word, and action may be judged.

This may shape behavior. But it also keeps morality at a lower level.

A mature person should not need eternal threat to understand that cruelty is wrong. A mature person should not need heavenly payment to understand that kindness is good. A mature person should not need divine surveillance to know that a child's suffering matters.

Real morality begins when we care about the effect of our actions on others.

Do not lie because lies damage trust. Do not steal because theft harms another person. Do not abuse power because the weak are vulnerable. Do not exploit the poor because desperation is not consent. Do not humiliate others because dignity matters. Do not harm children because they are defenseless and their wounds may last a lifetime.

These reasons do not require God. They require humanity.

A morality based on fear asks, "What will happen to me if I do this?"

A deeper morality asks, "What will happen to others if I do this?"

That difference is enormous.

Why Divine Surveillance Is Not Moral Maturity

Imagine a man who says, "I would harm others if I did not believe God was watching me."

Should we admire his faith?

Or should we worry about his character?

If the only barrier between a person and cruelty is the belief that he is being watched, then the problem is not atheism. The problem is that his moral life depends on surveillance.

Divine surveillance may make people behave better in some situations, just as cameras may reduce theft in a store. But being watched is not the same as being good.

A camera can change behavior without changing character. A police officer on the road can make a driver slow down without making him responsible. A strict parent can make a child obey without teaching wisdom.

In the same way, fear of God may restrain some behavior without creating deep compassion.

Mature morality means internalizing concern for others. It means understanding consequences. It means developing empathy. It means feeling responsibility even when no one is watching. It means doing the right thing not because a cosmic authority threatens punishment, but because the right thing matters.

This is why the accusation against atheists is morally backward.

Atheists do not believe in divine surveillance. If they act kindly, honestly, and responsibly, they do so without expecting heavenly reward and without fearing hell.

That does not make their morality weaker.

In some ways, it may make it cleaner.

A person who helps without expecting heaven shows that goodness can stand on its own. A person who tells the truth without fear of divine punishment shows that honesty can be human. A person who rejects cruelty without belief in hell shows that compassion does not need supernatural enforcement.

Religion often asks, "Why be good if God is not watching?"

A better question is: why are you not good unless God is watching?

Atheists Love, Care, Sacrifice, and Protect

The claim that atheists cannot be moral collapses immediately when we look at real life.

Atheists love their children. They care for aging parents. They visit sick friends. They donate blood. They serve as doctors, nurses, teachers, firefighters, soldiers, scientists, social workers, and caregivers. They comfort the grieving. They oppose slavery, racism, caste oppression, misogyny, cruelty, and injustice. They keep promises. They rescue strangers. They adopt children. They volunteer. They grieve when others suffer.

Atheists can be faithful spouses, devoted parents, honest workers, loyal friends, and courageous citizens.

This is not theory. It is ordinary human life.

Of course, atheists can also be selfish, cruel, dishonest, arrogant, or violent. Atheism does not make a person automatically good. No worldview does.

But the point is simple: atheists can be moral because moral capacity does not require belief in God.

Morality grows from human nature, social life, personal character, education, empathy, law, culture, reason, and shared vulnerability. Belief may influence morality, but it is not the source of morality.

A religious person may say, “But atheists borrow morality from religion.”

This is not convincing.

When an atheist says that cruelty is wrong because it causes suffering, that is not borrowed theology. It is a direct moral recognition. When an atheist says that women deserve equal dignity, that is not borrowed from ancient patriarchy. It is often a correction of religion. When an atheist says that slavery is wrong, that is not borrowed from scriptures that tolerated or regulated slavery. It is a moral judgment made against such traditions.

When an atheist says that caste hierarchy is unjust, that is not borrowed from sacred systems that protected hierarchy. It is a rejection of sacred inequality. When an atheist says that children should not be terrified with hell, that is not borrowed from religion. It is moral concern for the child's mind.

Atheists do not need to borrow morality from religion. They can build morality from the reality of harm, dignity, freedom, fairness, compassion, and human well-being.

The Real Question Is Human Suffering

Religious morality often asks, "What does God command?"

Secular morality asks a different question: who is harmed?

This is a better beginning.

A child beaten in the name of discipline is harmed. A girl denied education is harmed. A woman forced into submission is harmed. A lower caste person humiliated by sacred hierarchy is harmed. A teenager taught to hate his own honest questions is harmed. A doubter threatened with hell is harmed. A blasphemer punished for words is harmed. A patient denied medical care because of religious dogma is harmed. A community divided by sectarian hatred is harmed.

These harms matter whether or not a scripture approves them. They matter whether or not a priest defends them.

They matter whether or not tradition normalizes them. They matter because conscious beings suffer.

That is the foundation.

If morality begins with God's command, then suffering can be excused whenever religion calls it holy. If morality begins with human suffering, then no command is above examination.

This is the great strength of secular morality.

It does not ask first whether an action is ancient. It asks whether it is harmful. It does not ask first whether a rule is sacred. It asks whether it is just. It does not ask first whether authority approves. It asks whether dignity is protected.

This does not make moral questions easy. People may disagree about harm, rights, freedom, fairness, and consequences. Evidence may be incomplete. Values may conflict. Moral reasoning can be difficult.

But difficulty is not failure.

It is better to struggle honestly with human consequences than to obey ancient commands blindly.

A morality centered on suffering is more open to correction than a morality centered on divine authority.

That is why it can improve.

The Goodness of Ordinary People

Most human morality is not dramatic. It is ordinary.

A nurse adjusts a pillow for a dying patient. A father works long hours so his children can study. A neighbor brings food after surgery. A teacher stays after class to help a struggling student. A stranger returns a lost wallet. A friend listens patiently at midnight. A driver stops at a crosswalk. A doctor tells a hard truth gently. A daughter bathes an elderly mother. A passerby helps someone who has fallen.

These acts are the real substance of moral life.

Most of them do not require theological debate. They arise from seeing need and responding.

A person may be religious while doing them. A person may be atheist while doing them. The goodness is in the care.

Religion often tries to claim this ordinary goodness for God. But ordinary goodness is older and wider than religion. It appears wherever human beings recognize another person's need.

When a believer and an atheist both comfort a crying child, the child does not receive two different kinds of comfort. When a believer and an atheist both donate blood, the blood does not carry a religious label. When a believer and an atheist both tell the truth, truth is not different in their mouths. When a believer and an atheist both oppose cruelty, the moral value lies in opposing cruelty.

This should be obvious. But religion has made it seem controversial.

The world does not need more fear-based morality. It needs more ordinary decency: more honesty, more tenderness, more fairness, more courage, and more people willing to reduce suffering without asking whether the suffering person belongs to their group.

Why Goodness Does Not Need Heaven or Hell

Heaven and hell are often presented as moral necessities. Heaven rewards good. Hell punishes evil. God guarantees justice.

At first, this sounds comforting. In a world where many guilty people escape punishment and many innocent people suffer, divine justice can feel emotionally satisfying.

But there is a problem.

A morality based on heaven and hell can become self-centered. If I help others mainly because I want heaven, my action is partly self-serving. If I avoid cruelty mainly because I fear hell, my restraint is based on self-protection. If I obey religious rules mainly to secure my own eternal future, morality becomes a transaction between my soul and God.

This is not the highest moral vision.

A better morality asks us to care about others for their sake.

Feed the hungry because hunger hurts. Protect children because children are vulnerable. Defend freedom because oppression damages life. Tell the truth because trust matters. Support justice because victims matter. Care for the sick because suffering should be relieved.

No heaven is needed. No hell is needed.

The moral reason is already present in the human situation.

This does not mean consequences are irrelevant. Human societies need laws, accountability, rewards, and penalties. But these are human tools for protecting life and reducing harm. They do not require eternal torture or heavenly reward.

In fact, the idea of hell creates its own moral problem.

Infinite punishment for finite human wrongdoing is morally horrifying. Eternal torture for honest disbelief is not justice. It is cruelty magnified beyond imagination.

A religion that threatens eternal punishment for honest doubt does not stand on moral high ground.

It stands on fear.

A humane morality can do better.

Motivation Is Not Ownership

Religion may motivate some people to behave well. That should be admitted.

For some believers, faith encourages charity, forgiveness, patience, discipline, service, and sacrifice. Religious communities may organize food drives, hospitals, shelters, schools, and disaster relief. Religious rituals may strengthen gratitude, humility, and belonging.

These are real human benefits.

But motivation is not the same as moral truth.

A person may be motivated to do good by religion, but the goodness of the act does not depend on religion. If a church feeds hungry families, feeding them is good because hunger is relieved. If a mosque helps refugees, helping them is good because vulnerable people are protected. If a temple serves free meals, serving them is good because people need food. If an atheist organization does the same, the moral value is the same.

The hungry stomach does not care whether food arrived through prayer, humanism, friendship, or public policy. The relief is real. The moral truth is in the reduction of suffering.

We should not confuse the source of motivation with the foundation of morality.

Religion can motivate good behavior. So can empathy, humanism, love, reason, personal experience of suffering, education, friendship, moral imagination, and the simple recognition that another being matters.

If religion were the only possible moral motivation, atheists would be incapable of goodness.

But they are not.

That fact alone disproves the claim.

The Problem of Selective Religious Morality

Religion often makes morality selective.

Love your neighbor, but define the neighbor narrowly. Protect your own community, but suspect outsiders. Care for believers, but condemn unbelievers. Honor life, but justify holy violence. Defend family, but control women. Preach compassion, but threaten hell. Praise humility, but claim exclusive truth. Condemn cruelty, but frighten children with eternal punishment.

This selective morality is common because religions are tied to identity, tribe, authority, and tradition.

When morality is filtered through religious loyalty, the outsider becomes morally less important. The unbeliever becomes dangerous. The apostate becomes a traitor. The critic becomes evil. The person of another faith becomes lost, impure, or damned.

This is not universal morality. It is group loyalty dressed as divine truth.

A secular morality can also become tribal. Atheists are not immune to arrogance, ideology, or group hatred. Any worldview can become dogmatic if it refuses criticism.

But science-based atheism has an important advantage when it remains true to itself: it does not need sacred boundaries between human beings. It does not need to divide people into saved and damned. It does not need to call doubt a sin. It does not need to make one book, one prophet, one priesthood, or one tradition immune from criticism.

It can say simply: this person can suffer; that suffering matters; this person has dignity; that dignity matters; this person's freedom should not depend on sharing my belief.

That is a broader moral starting point than religious tribalism.

Are Religious People More Moral?

This question must be handled carefully.

It would be dishonest to say that religion never encourages moral behavior. It often does. It can build community, encourage charity, create accountability, provide rituals of gratitude, and inspire service.

It would also be dishonest to say that religion reliably makes people morally superior. History and daily life do not support that.

Religious people can be kind, cruel, generous, selfish, peaceful, violent, honest, corrupt, humble, arrogant, compassionate, or hateful. Atheists can be the same.

Belief or disbelief alone does not determine character.

The relationship between religion and morality is complicated. It depends on what kind of religion, what kind of society, what kind of behavior, what kind of measurement, and whether morality is directed toward insiders or outsiders.

Religion may increase generosity toward members of the same group while increasing suspicion toward outsiders. It may encourage charity while also reinforcing prejudice. It may comfort suffering while also creating fear. It may teach kindness while also teaching obedience to harmful authority. It may support moral behavior in some contexts and immoral behavior in others.

This mixed picture is exactly what we would expect if morality does not come directly from God.

Religion is a human institution. Like all human institutions, it can bring out good or bad depending on its teachings, leaders, culture, and use of power.

The better question is not, "Are religious people moral?"

Many are.

The better question is, "Does morality require religion?"

It does not.

The Moral Strength of Atheism

Atheism by itself is not a moral system. It simply means lack of belief in God.

That must be said clearly.

A person can be an atheist and still be selfish, cruel, dishonest, or indifferent. Rejecting God does not automatically produce wisdom.

But atheism can clear the ground for a better morality when it is joined with humanism, reason, compassion, science, and respect for human dignity.

It removes divine excuses.

No one can say, “God commanded this cruelty.” No one can say, “This hierarchy is sacred.” No one can say, “This child deserves fear because scripture says so.” No one can say, “This woman must submit because God made it so.” No one can say, “This caste is lower by divine order.” No one can say, “This unbeliever deserves punishment because God is offended.”

Science-based atheism asks for human reasons.

If a rule causes harm, defend it with evidence. If a tradition humiliates people, explain why it should continue. If a belief frightens children, show that the fear is necessary. If a command restricts freedom, justify it in human terms. If a punishment is cruel, do not hide behind God.

This is morally powerful.

Atheism does not guarantee goodness. But it removes one of the oldest shields for cruelty: divine permission.

The Better Question

The religious question is: why be good without God?

The better question is: what kind of morality best protects conscious beings from unnecessary suffering?

Once we ask that question, the conversation changes. We stop asking whether a rule is sacred and ask whether it is harmful or helpful. We stop asking whether an authority commands it and ask whether it respects dignity. We stop asking whether a tradition is old and ask whether it is just. We stop asking whether God is offended and ask whether people are wounded.

That is moral progress.

A secular morality does not have all answers ready-made. It must reason, listen, study consequences, correct mistakes, and remain humble. But this is a strength, not a weakness.

A morality that can admit error can improve. A morality that claims divine perfection often cannot.

That is why atheism need not lead to moral emptiness. It can lead to moral responsibility.

If there is no God to judge us, then we must judge our actions more honestly. If there is no heaven to correct injustice later, then we must fight injustice now. If there is no hell for the cruel, then we must protect victims here. If there is no divine plan, then human choices matter more, not less.

Atheism does not say, "Nothing matters."

It says, "This world matters." It says suffering matters, freedom matters, truth matters, and human beings must stop waiting for heaven to do the work of justice.

The Beauty of Goodness Without God

There is something deeply beautiful about goodness without God.

A person helps another person, knowing there is no heavenly reward. A doctor cares for a dying patient, knowing there may be no afterlife. A stranger risks safety to save someone, not because God commands it, but because another life matters. A parent loves a child without imagining that love must be eternal to be real. A human being acts kindly in a vast universe, not because a divine judge is watching, but because kindness itself is worth doing.

That kind of goodness is not empty.

It is noble.

It is goodness standing on its own feet.

Religion often says that without God, morality loses its foundation. But perhaps the opposite is true. When morality no longer depends on reward, punishment, scripture, priesthood, or divine surveillance, it becomes more honest.

It asks us to grow up. It asks us to care without being bribed. It asks us to be responsible without being watched. It asks us to protect one another because no god will do it for us.

This is not moral collapse.

This is moral adulthood.

Chapter Summary

This chapter answered one of the most common religious accusations against atheism: without God, anything is permitted.

That accusation is false. Morality does not depend on belief in God. Human beings can be moral because they can understand suffering, feel empathy, build trust, recognize fairness, care for children, form relationships, reason about consequences, and take responsibility for one another.

Fear of hell is not the same as goodness. Hope of heaven is not the same as compassion. Divine surveillance is not moral maturity.

Atheists can love, care, sacrifice, protect, tell the truth, defend justice, and live responsibly without belief in God. Religion may motivate some moral behavior, but it does not own morality.

The real foundation of morality is not divine command. It is the reality that conscious beings can be harmed or helped.

A secular morality asks better questions: Who is suffering? What reduces harm? What protects dignity? What supports freedom, fairness, and human well-being?

Without God does not mean without goodness.

It means morality must stand honestly on human ground.

Key Takeaways

The claim that atheists cannot be moral is a prejudice, not a serious argument.

Fear of divine punishment may control behavior, but fear is not the highest form of morality.

Goodness does not require heaven, hell, or divine surveillance.

Atheists can be loving, honest, generous, responsible, and courageous.

Religion may motivate some good behavior, but it did not create morality and does not own goodness.

The moral value of an action depends on its effect on conscious beings, not on whether it is religiously labeled.

A secular morality can be deeper than fear-based morality because it begins with suffering, dignity, compassion, and responsibility.

Transition to the Next Chapter

If morality does not require God, then we must examine the next question.

What happens when morality is based on God?

Religion claims that divine command gives morality a firm foundation. But history shows a darker truth. Sacred commands have often been used to defend cruelty, slavery, caste, patriarchy, persecution, violence, and fear.

A command does not become moral because it is called holy.

A cruelty does not become kindness because it is written in scripture.

The next chapter will examine what happens when divine morality fails.

Sources

Phil Zuckerman, *Society Without God.*

Greg M. Epstein, *Good Without God.*

Frans de Waal, *The Bonobo and the Atheist.*

Michael Tomasello, *A Natural History of Human Morality.*

Richard Joyce, *The Evolution of Morality.*

R. McKay and H. Whitehouse, "Religion and Morality," *Psychological Bulletin*, 2015.

T. Ståhl, "The Amoral Atheist? A Cross-National Examination of Cultural, Motivational, and Cognitive Antecedents of Disbelief, and Their Implications for Morality," *PLOS ONE*, 2021.

W. M. Gervais et al., “Global Evidence of Extreme Intuitive Moral Prejudice Against Atheists,” *Nature Human Behaviour*, 2017.

T. Bendixen and others, “Gods Are Watching and So What? Moralistic Supernatural Punishment Across Cultures,” 2023.

L. van Ments et al., “Modelling the Effect of Religion on Human Empathy Based on Neurological Evidence,” 2018.

Chapter 9: When Divine Morality Fails

Obedience, Scripture, Cruelty, and the Problem of Holy Commands

A command does not become moral because it is called sacred. A cruelty does not become kindness because it is written in scripture. A chain does not become freedom because religion blesses it. A wound does not stop bleeding because a priest, prophet, or holy book says the wound was deserved.

This is one of the greatest moral lessons humanity must learn.

Religion often claims that morality needs God. It says human beings are too weak, too selfish, too sinful, and too confused to know good from evil without divine command. It says God provides the moral law, scripture preserves it, religious authority teaches it, and obedience protects society from moral collapse.

At first, this sounds reassuring.

But history tells a darker story.

The same religious authority that says "love your neighbor" has also been used to defend slavery. The same scripture that speaks of compassion has also been used to justify patriarchy. The same faith that praises mercy has also punished doubt. The same God who is called loving has been invoked to defend holy war, caste hierarchy, persecution, forced conversion, blasphemy laws, and fear of hell.

This does not mean every religious person is cruel. That would be false and unfair. Many religious people are kind, generous, honest, loving, and deeply committed to service. They feed the hungry, care for the sick, comfort the grieving, and protect the vulnerable.

But when they do so, they are often following their humanity more than the harshest parts of their religious tradition. They choose compassion over cruelty. They choose conscience over literal command. They choose human dignity over ancient text.

And that is exactly the point.

If moral judgment must soften, correct, reinterpret, or reject sacred commands, then morality does not truly come from those commands. Something deeper is at work.

Human conscience must stand above scripture. Human suffering must matter more than religious obedience. Human dignity must be stronger than holy authority.

This chapter examines what happens when morality is placed under divine command. It shows why obedience is a dangerous foundation for ethics.

Is It Good Because God Commands It?

One of the oldest and strongest questions in moral philosophy comes from Plato's dialogue *Euthyphro*. The question can be stated simply: is something good because God commands it, or does God command it because it is good?

This may sound like a philosopher's puzzle, but it is not. It is one of the most important moral questions ever asked.

If something is good only because God commands it, then morality becomes obedience to power. In that case, anything could become moral if God commanded it.

Slavery could become moral. Genocide could become moral. Stoning could become moral. The oppression of women could become moral. The humiliation of a lower caste person could become moral. The killing of unbelievers could become moral. The punishment of children for doubt could become moral.

All that would be needed is a divine command.

That is terrifying.

A command alone cannot make cruelty good. A command alone cannot make injustice holy. A command alone cannot turn suffering into righteousness.

But if God commands something because it is already good, then goodness does not depend on God. It exists as a standard that even a divine command must respect. In that case, we must still use moral judgment to recognize goodness.

Either way, divine-command morality faces a serious problem. If morality depends only on God's command, it becomes arbitrary and potentially cruel. If God commands what is good because it is good, then morality is not dependent on God.

This is the heart of the Euthyphro dilemma.

Most decent believers understand this instinctively. If asked, "Would torture become good if God commanded it?" many would say no. If asked, "Would slavery become moral if God approved it?" many would say no. If asked, "Would killing a child become righteous if a holy text demanded it?" many would say no.

That answer is morally important. It means they are judging the command by a standard deeper than the command itself. It means conscience is still alive.

It means goodness is not the same as obedience.

A holy order is not automatically a moral order.

When Obedience Pretends to Be Morality

Obedience is useful in some parts of life. A child must learn not to run into traffic. A student must follow certain rules in school. A soldier may need discipline in danger. A society needs laws.

But obedience is not the highest form of morality. In fact, obedience can become dangerous when it is separated from conscience.

A person who only asks, "What am I commanded to do?" may stop asking, "Is this right?" A person who only asks, "What does my scripture say?" may stop asking, "Who is being harmed?" A person who only asks, "What does my religious authority teach?" may stop asking, "Does this protect dignity?"

This is how good people can participate in bad systems.

They do not always begin as monsters. They begin as obedient believers, loyal citizens, faithful followers, respectful children, or disciplined members of a group. They are told that obedience is virtue. They are told that doubt is pride. They are told that questioning is rebellion. They are told that human compassion must submit to divine law.

Then conscience becomes weaker.

The person may feel pity but suppress it. The person may see harm but justify it. The person may hear suffering but call it necessary. The person may feel moral discomfort but silence it because a sacred authority has spoken.

That is the danger.

Divine morality can train people to distrust their own compassion.

A father may reject his child because religion condemns the child's sexuality. A mother may teach her daughter submission because scripture says women must obey. A community may humiliate a lower caste person because tradition calls the hierarchy sacred. A believer may support blasphemy punishment because religion says God must be defended. A child may be frightened with hell because adults call it spiritual truth.

In each case, obedience replaces conscience. And when obedience replaces conscience, morality has already begun to fail.

Scripture Is Not a Clear Moral Guide

Religious morality is often presented as clear, eternal, and unchanging. But real religious traditions are full of moral contradictions.

One passage may speak of mercy. Another may command punishment. One tradition may praise compassion. Another may defend exclusion. One teacher may preach love. Another may demand obedience. One believer may say religion supports peace. Another may use religion to justify violence. One community may interpret scripture symbolically. Another may enforce it literally.

This is why religious morality is not as simple as believers often claim.

If scripture were a clear and perfect moral guide, sincere believers would not disagree so deeply about slavery, war, women's rights, caste, sexuality, punishment, democracy, freedom of speech, religious liberty, and the rights of unbelievers.

But they do disagree.

They disagree because they are not simply reading morality from scripture. They are interpreting scripture through culture, history, personality, politics, power, education, and existing moral instincts.

When society improves, many believers reinterpret their texts. When slavery becomes morally unacceptable, they discover that their religion was truly against slavery all along. When women demand equal dignity, they reinterpret old patriarchal passages. When caste discrimination becomes indefensible, they soften or reject religious justifications. When religious violence becomes embarrassing, they emphasize peaceful verses. When science defeats old beliefs, they call the defeated beliefs symbolic.

This process reveals something important.

Scripture is often not leading moral progress. It is being dragged by moral progress.

Human conscience moves forward, and religion later adjusts.

This does not mean every religious reform is insincere. Many reformers are brave and deeply moral. But their bravery often comes from challenging inherited religious authority, not merely obeying it.

If a holy book can be used to defend both slavery and abolition, both patriarchy and equality, both violence and peace, then the book itself is not enough.

Human moral judgment must decide which interpretation is acceptable.

Again, morality stands above scripture.

Slavery and Sacred Permission

Slavery is one of the clearest examples of divine morality failing.

For centuries, human beings owned other human beings. They bought them, sold them, whipped them, separated families, forced labor, and treated persons as property.

This was not defended only by greed. It was also often defended by religion.

Slaveholders and their defenders found verses, traditions, and theological arguments to support slavery. They said slavery was permitted by God, regulated by scripture, accepted by prophets, or necessary for social order. Enslaved people were told to obey. Masters were told they had divine authority. Cruelty was wrapped in religious language.

Today, most decent people see slavery as evil.

But that moral clarity did not come simply from obeying ancient scripture. It came from recognizing the humanity and suffering of enslaved people. It came from moral progress. It came from human beings saying: no scripture can make this right.

This is the key point.

If scripture permits or regulates slavery, and human conscience later says slavery is wrong, then conscience has corrected scripture. If believers must reinterpret scripture to reject slavery, then morality is not simply coming from scripture. It is coming from a deeper recognition of human dignity.

No divine command can make one human being the property of another. No sacred text can make the slave's pain disappear. No priestly explanation can turn chains into justice.

The suffering of the enslaved person is morally louder than the permission of any scripture.

Caste and Sacred Inequality

Caste hierarchy offers another painful example.

In caste-based societies, human beings have been ranked by birth. Dignity has been distributed unequally. Humiliation has been normalized. Social position has been treated as destiny. People born into oppressed castes have been told, directly or indirectly, that their suffering belongs to sacred order.

Religious ideas helped make this hierarchy appear natural, cosmic, and morally acceptable.

This is divine morality failing.

No birth can make a person less human. No ritual status can erase dignity. No sacred story can justify humiliation. No priestly system can make inequality moral by calling it holy.

If a child is born into a group treated as polluted or inferior, the moral issue is not that the child has a low spiritual status. The moral issue is that society has committed an injustice against that child before the child has even spoken a word.

The child's dignity comes before the ritual label. The human being comes before the hierarchy.

If a religious system says some people are born higher and others lower, the moral response must be clear: the system is wrong.

This is why morality must stand above sacred authority. Without that courage, old injustice can wear holy clothing for centuries.

Patriarchy and the Sacred Control of Women

Patriarchy, too, has often been protected by religion.

Women have been told to obey husbands, remain silent, accept male authority, cover their bodies, limit their ambitions, sacrifice their freedom, and treat submission as virtue. Religious traditions have often controlled women's sexuality, education, mobility, inheritance, leadership, and public voice.

This control is often presented as protection, purity, modesty, family order, or divine design.

But we must ask the moral question directly. Who benefits? Who is controlled? Whose voice is silenced? Whose body is regulated? Whose education is limited? Whose freedom is sacrificed?

A rule does not become moral because it is called modesty. A hierarchy does not become just because it is called family order. Submission does not become dignity because it is described as divine plan.

A woman is not a lesser moral being. She is not a supporting character in a man's spiritual story. She is not property, temptation, servant, reproductive vessel, or symbol of family honor.

She is a full human being.

If a religious command denies that, the command is wrong.

Again, many believers today reject crude patriarchy. That is good. But the rejection usually comes because human moral understanding has improved, not because ancient religious texts were always clear in favor of equality.

Human dignity corrected religious authority.

That is moral progress.

Persecution in the Name of God

Religion becomes especially dangerous when it treats disagreement as evil.

If a political leader is criticized, he may become angry. If a scientific theory is challenged, it can be tested. If a human law is criticized, it can be debated. But if a belief is treated as divine truth, criticism can become blasphemy. And blasphemy can become a crime.

This is one of the great dangers of sacred authority.

The unbeliever becomes not merely mistaken, but wicked. The apostate becomes not merely independent, but traitorous. The heretic becomes not merely different, but poisonous. The critic becomes not merely skeptical, but offensive to God.

Once this happens, cruelty becomes easy. Silence the critic. Punish the doubter. Expel the heretic. Threaten the apostate. Censor the writer. Ban the book. Control the child. Protect God's honor.

Of course, God should not need human protection. An all-powerful being should not be injured by a sentence, cartoon, question, or book.

But religious authority often needs protection.

That is the real issue.

Blasphemy laws and persecution often protect religious power from examination. They teach people to fear thought. They make honest doubt dangerous. They turn fragile ideas into untouchable commands.

A society cannot be morally healthy if questions are treated as crimes. A person cannot be intellectually free if doubt is treated as sin. A child cannot grow honestly if fear of hell is placed before curiosity.

Persecution in the name of God is not a defense of morality. It is a confession that the belief cannot survive freedom.

The Cruelty of Sacred Fear

Not all religious harm is physical. Some of it happens inside the mind.

A child is told that God watches every thought. A teenager is told that normal desire makes the soul dirty. A doubter is told that honest questions may lead to eternal punishment. A grieving person is told that only believers will see loved ones again. A woman is told that disobedience makes her sinful. A lower caste child is told that humiliation is part of sacred order. A nonbeliever is told that goodness without God is impossible.

Fear enters early. It enters before reason is mature. It enters before the child can examine evidence. It enters before the mind has defenses.

This fear can last for decades.

Some adults who no longer believe still feel fear of hell. Some still feel shame about the body. Some still feel guilt for questioning. Some still feel anxiety when they reject a religious command. Some still hear old voices inside them saying, “You are sinful, impure, rebellious, lost.”

This is not moral education.

It is psychological control.

A morality that depends on frightening children with eternal punishment should not be considered morally superior. A morality that teaches people to fear their own thoughts has already damaged human freedom. A morality that places obedience above mental health is not sacred.

It is harmful.

The question is simple: does a belief help human beings flourish, or does it wound them?

If it wounds them, calling it holy does not heal the wound.

Why Moral Progress Often Requires Disobedience

Many of humanity's greatest moral advances required disobedience.

Disobedience to kings, priests, caste rules, slaveholders, patriarchal authority, segregation, unjust laws, and sacred traditions.

The abolitionist had to disobey the moral comfort of slaveholding society. The woman demanding education had to disobey those who said her place was silence. The lower caste reformer had to disobey the religious order that humiliated him. The scientist had to disobey the doctrine that made scripture superior to evidence. The atheist had to disobey the command to believe without proof.

The moral reformer often begins as a troublemaker. Only later does society call him brave.

This is why obedience cannot be the highest virtue. If obedience were always moral, no unjust system would ever be challenged.

A person must sometimes say: this command is wrong, this scripture is wrong, this tradition is wrong, this priest is wrong, this prophet is wrong, this god, if he commands cruelty, is wrong.

That last sentence is the true test of moral courage.

If a person cannot say that a divine command would be wrong if it commanded cruelty, then that person has surrendered conscience. But if a person can say it, then morality does not come from divine command.

It comes from a standard deeper than obedience.

The Courage to Judge Holy Claims by Human Consequences

Every moral claim should be judged by its consequences for conscious beings.

Does it reduce suffering or increase it? Does it protect dignity or violate it? Does it expand freedom or crush it? Does it encourage honesty or punish questions? Does it protect children or frighten them? Does it honor women or control them? Does it respect the poor or blame them?

Does it defend equality or sanctify hierarchy? Does it make people kinder, wiser, freer, and more responsible? Or does it make them fearful, obedient, tribal, ashamed, and cruel?

These questions are better than asking, "Is it written?"

Many things are written. The important question is whether they are true, just, and humane.

A sacred text is still a text. A tradition is still a tradition. A command is still a command. A moral claim must earn respect. It cannot demand respect merely by calling itself divine.

This is the heart of secular morality.

Nothing is above criticism: no book, no prophet, no priest, no tradition, no god.

If a command causes unnecessary suffering, we must question it. If a belief humiliates people, we must question it. If a tradition protects injustice, we must question it. If a scripture defends cruelty, we must question it.

This is not arrogance. It is responsibility.

If no divine authority has proven itself beyond doubt, then human beings must take responsibility for moral judgment.

We cannot hide behind God.

Religious People and Moral Selection

Most modern religious believers do not follow every ancient command literally.

That is good.

They select. They emphasize kindness. They soften harsh rules. They ignore cruel passages. They reinterpret old laws. They say certain commands belonged to another time. They focus on love, mercy, charity, and forgiveness.

Again, this is good.

But notice what is happening. They are using moral judgment to decide which parts of religion are acceptable. They are not getting morality simply from scripture. They are bringing morality to scripture.

When a believer says, "That passage should not be followed today," he is using a standard outside the passage. When a believer says, "This command must be understood metaphorically," she is using moral judgment. When a believer says, "God would not really want cruelty," he is judging the command by an idea of goodness that is not identical to the command itself.

This moral selection is unavoidable.

It is also revealing.

The best religious people are often better than their scriptures. They are more compassionate than the harshest verses. They are more humane than ancient punishments. They are more equal-minded than patriarchal laws. They are more tolerant than exclusivist doctrines. They are more loving than threats of hell.

Their goodness comes from human moral growth.

Religion may give them language for that goodness, but it does not create the goodness.

In many cases, humanity must rescue morality from religion.

The Problem With Saying "God Is Good"

Some believers try to escape the Euthyphro dilemma by saying, "God's nature is goodness itself."

This sounds elegant, but it does not solve the problem.

How do we know God is good?

If we say God is good because whatever God does is good, then the word "good" loses meaning. It only means "whatever God is or does." If God commands mercy, mercy is good. If God commands slaughter, slaughter becomes good. If God blesses equality, equality is good. If God creates hierarchy, hierarchy becomes good.

In that case, goodness becomes empty praise for power.

But if we say God is good because God is loving, just, compassionate, truthful, and merciful, then we are using moral standards to evaluate God. We already know that love is better than cruelty, justice is better than injustice, compassion is better than indifference, truth is better than deception, and mercy is better than sadism.

Those standards do not come merely from the statement "God is good."

We bring them to the statement.

The phrase "God is good" only makes sense if we already have some understanding of goodness.

This is why divine morality cannot escape human moral judgment. A person who says "God is good" must still explain what goodness means.

And once we ask what goodness means, we are back to human realities: suffering, compassion, dignity, honesty, justice, freedom, fairness, and well-being.

God is no longer necessary.

What Divine Morality Cannot Fix

Divine morality claims to give certainty.

But certainty is not the same as truth.

A person can be certain and wrong. A community can be certain and cruel. A scripture can be treated as certain and still contain moral error.

Certainty can even make cruelty worse because it removes doubt.

A person who is uncertain may hesitate before harming others. A person who thinks God commands harm may act with frightening confidence.

This is one of the dangers of religious violence.

Ordinary cruelty knows it is cruelty. Sacred cruelty thinks it is duty.

That makes it harder to stop.

A person who harms others for greed may be ashamed if exposed. A person who harms others for God may feel proud.

This is why divine morality can be more dangerous than ordinary selfishness. It can make people feel righteous while doing harm. It can make compassion seem like weakness. It can make doubt seem like betrayal. It can make mercy seem like disobedience. It can make cruelty feel clean.

No moral system should have that power.

A Better Foundation Than Command

If divine command fails, what should replace it?

The answer is not moral chaos. The answer is moral responsibility.

Begin with conscious beings. They can suffer. They can flourish. They can be frightened, humiliated, enslaved, loved, educated, healed, wounded, silenced, or freed.

That is where morality begins.

Then ask: what reduces unnecessary suffering? What protects dignity? What respects freedom? What improves well-being? What treats people fairly? What protects children? What supports truth? What allows correction when we are wrong?

This foundation is not perfect because human beings are not perfect. But it is better than obedience to commands that cannot be questioned.

A morality based on human well-being can learn. A morality based on evidence can improve. A morality based on compassion can expand. A morality based on freedom can protect dissent. A morality based on dignity can challenge hierarchy. A morality based on humility can admit mistakes.

Divine-command morality often says, "Obey."

Human morality must say, "Think, care, examine, and take responsibility."

That is harder.

But it is better.

Chapter Summary

This chapter examined the failure of divine morality.

Religion often claims that morality comes from God's command. But this creates a serious problem. Is something good because God commands it, or does God command it because it is good? If something is good only because God commands it, then morality becomes obedience to power. If God commands something because it is already good, then goodness exists independently of God.

Divine-command morality can become dangerous because it may train people to value obedience over conscience. Sacred authority has often been used to defend slavery, caste hierarchy, patriarchy, persecution, blasphemy laws, fear of hell, and punishment of doubt.

This does not mean religious people cannot be moral. Many are deeply moral. But when they reject cruel religious commands, reinterpret harsh scriptures, and choose compassion over obedience, they show that moral judgment stands above scripture.

A better morality begins with conscious beings who can suffer and flourish. It asks what reduces harm, protects dignity, supports freedom, respects fairness, and allows moral correction.

A command does not become moral because it is called holy.

Key Takeaways

Divine-command morality faces the Euthyphro problem: is something good because God commands it, or does God command it because it is good?

If morality depends only on divine command, then cruelty could become good if commanded by God.

If God commands what is already good, then morality does not depend on God.

Obedience is not the highest moral virtue. Conscience must be able to judge commands.

Religious traditions have often been used to defend slavery, caste, patriarchy, persecution, and fear-based control.

Many good religious people are moral because they choose compassion over the harshest parts of their traditions.

No scripture, prophet, priest, tradition, or god should be placed above human suffering and dignity.

Transition to the Next Chapter

If divine morality fails, then we need a better foundation.

That foundation is not moral emptiness. It is not selfishness. It is not chaos.

It is the reality of conscious life.

Human beings can suffer. They can be harmed. They can be humiliated. They can be oppressed. They can also be protected, healed, educated, loved, and freed.

The next chapter will build the positive alternative: a better morality for a real world, based on harm, dignity, justice, freedom, and human well-being.

Sources

Plato, *Euthyphro.*

Mark C. Murphy, "Theological Voluntarism," *Stanford Encyclopedia of Philosophy.*

Stanford Encyclopedia of Philosophy, "Religion and Morality."

Internet Encyclopedia of Philosophy, "Divine Command Theory."

H. Van Eyghen, "Debunking Divine Command Theory," *Religions*, 2023.

Richard Joyce, *The Evolution of Morality.*

Peter Singer, *Practical Ethics.*

Bertrand Russell, "Why I Am Not a Christian."

A.C. Grayling, *The God Argument.*

Robert McKay and Harvey Whitehouse, "Religion and Morality," *Psychological Bulletin*, 2015.

Chapter 10: A Better Morality for a Real World

Harm, Dignity, Freedom, Justice, and Human Well-Being

A hungry child does not need a sermon. A beaten woman does not need a verse. A persecuted minority does not need a holy argument. A frightened child does not need a doctrine of sin. A patient in pain does not need a metaphysical explanation.

They need food, safety, dignity, freedom, care, protection, and relief from suffering.

That is where morality should begin.

Not in heaven. Not in ancient command. Not in the fear of hell. Not in obedience to priests, prophets, scriptures, or tradition.

Morality should begin with the real condition of conscious beings in the real world.

Can they suffer? Can they be harmed? Can they be humiliated? Can they be enslaved? Can they be silenced? Can they be healed? Can they be educated? Can they be loved? Can they be freed? Can they flourish?

These questions are better than asking, "What does God command?"

A morality worthy of human beings must begin where human life actually happens: in the body, in the mind, in the home, in the hospital, in the school, in the workplace, in the street, in the court, in the family, in the prison, in the village, in the nation, and in the fragile inner life of each person.

In the previous chapter, we saw why divine-command morality fails. A command does not become moral because it is called holy. A cruelty does not become kindness because it is written in scripture. If a belief causes unnecessary suffering, humiliation, fear, or oppression, it must be questioned, no matter how sacred it claims to be.

But criticism is not enough.

It is not enough to say that divine morality fails. We must ask what should replace it.

The answer is not chaos. It is not selfishness. It is not "anything goes."

The answer is a more honest, more human, more compassionate morality built on harm, dignity, freedom, justice, and well-being.

This morality does not need God. It needs courage, evidence, compassion, reason, and human beings willing to take responsibility for one another.

Begin With Suffering

Suffering is not imaginary.

Pain hurts. Hunger weakens. Fear shrinks the mind. Humiliation wounds dignity. Loneliness breaks the spirit. Violence damages the body and memory. Oppression steals possibility. Abuse can echo through a lifetime.

A child terrified by hell is not suffering in a symbolic way. The fear is real. A woman trapped in religious submission is not oppressed in a theoretical way. Her freedom is real. A lower caste person humiliated by sacred hierarchy is not injured in an abstract way. The humiliation is real. A young person taught that honest doubt is sin is not merely having a religious disagreement. His mind is being trained to fear truth.

A patient denied care because of religious dogma does not suffer as an idea. The body suffers. A person punished for blasphemy, apostasy, or doubt is not harmed in some distant philosophical sense. The harm happens in life.

This is why a better morality must begin with suffering.

If a belief increases suffering, we must examine it. If a tradition protects suffering, we must question it. If a law causes suffering, we must challenge it. If a sacred command excuses suffering, we must refuse to obey it.

This does not mean that every painful thing is immoral. Surgery can hurt but heal. Discipline can be uncomfortable but necessary. Hard truth can wound pride but protect life. Justice may require consequences. Parenting, medicine, education, and law sometimes require difficult decisions.

But avoidable suffering matters. Unnecessary cruelty matters. Preventable harm matters.

A morality that ignores suffering because a command is sacred is not moral at all.

Religion often begins with God's honor. A better morality begins with human pain.

Religion often asks whether a rule has divine approval. A better morality asks whether the rule harms or helps conscious beings.

Religion often worries about sin. A better morality worries about suffering.

That is not a lower morality.

It is a more serious one.

Harm Is Not a Matter of Theology

If a child is burned, the pain does not depend on the child's religion. If a woman is beaten, the injury does not depend on scripture. If a man is enslaved, his suffering is not reduced because slavery was once blessed by holy authority. If a person is humiliated by caste, the wound is not healed because the hierarchy is called sacred. If a dissenter is imprisoned, the loss of freedom does not become moral because the state claims to defend God.

Harm is real before theology speaks.

This is one of the strongest arguments for secular morality. A person does not need to settle the existence of God before knowing that torture is wrong. A society does not need to agree on heaven before protecting children from abuse. A court does not need scripture before recognizing that murder, rape, slavery, and exploitation destroy human life. A doctor does not need divine command before treating pain. A parent does not need a priest before comforting a frightened child.

Human beings can recognize harm because we are vulnerable beings ourselves. We know pain from inside. We know fear from inside. We know grief from inside. We know humiliation from inside. We know what it means to be ignored, threatened, used, or betrayed.

This shared vulnerability is one of the deepest sources of morality.

When I understand that you can suffer as I can suffer, morality begins. When I understand that your child's pain matters as my child's pain matters, morality expands. When I understand that a stranger's dignity does not depend on my religion, morality becomes universal.

This does not require God. It requires moral imagination. It requires the ability to say, "That could be me. That could be my child. That could be my mother. That could be my future."

Religious morality often divides the world into believers and unbelievers, pure and impure, saved and damned, chosen and rejected.

A morality based on harm crosses those boundaries. It says suffering matters wherever it occurs.

That is a better beginning.

Dignity Before Doctrine

Human dignity means that a person has worth before any label is added.

Before religion. Before caste. Before gender. Before nationality. Before wealth. Before social status. Before education. Before usefulness. Before obedience. Before belief.

A newborn has dignity before he can pray. A woman has dignity whether or not a scripture grants it. A poor man has dignity whether or not society respects him. An atheist has dignity even if religion condemns him. A prisoner has dignity even after wrongdoing. A person with dementia has dignity even when memory fades. A disabled person has dignity even when independence is limited. A dying patient has dignity even when cure is impossible.

This idea is morally powerful because it does not make human worth depend on divine approval. It does not say that dignity belongs only to one faith, one caste, one sex, one tribe, one race, one nation, or one moral category.

It says that human beings matter because they are conscious, vulnerable, feeling beings with lives that can go better or worse.

This does not mean all actions are equal. A murderer and a victim are not morally the same in action. A cruel person and a kind person are not morally the same in character. Justice must still distinguish harm from help, cruelty from compassion, guilt from innocence.

But even when a person does wrong, the person does not become an object.

A humane society can punish without torture. It can restrain without degrading. It can protect victims without becoming cruel. It can hold people accountable without pretending that some human beings have no dignity at all.

This is where secular morality can be stronger than divine morality.

Divine morality may say: this person is sinful, impure, damned, infidel, heretic, lower-born, cursed, or spiritually polluted.

Human dignity says: this person is still a person.

That is the moral line we must not cross.

Defend Freedom

Freedom is not just a political slogan. Freedom is part of human well-being.

A person's life is lived from the inside. Each person has fears, hopes, memories, loves, abilities, needs, and possibilities. To crush freedom is not merely to control behavior. It is to shrink a life.

A woman denied education loses more than school. She loses future choices. A child forbidden to ask questions loses more than information. He loses intellectual courage. A person forced into a religion loses more than belief. She loses inner honesty. A lower caste person denied dignity loses more than status. He loses equal standing in the human community. A writer censored for blasphemy loses more than publication. He loses freedom of thought. A girl forced into marriage loses more than childhood. She loses ownership of her own life. A person threatened for leaving religion loses more than community. He loses the right to conscience.

This is why freedom matters morally.

Religion often praises obedience. But obedience is not the same as goodness. A person forced to obey has not become moral. A person frightened into silence has not become virtuous. A person kept ignorant has not become pure.

A better morality values freedom because human beings must be able to think, question, choose, learn, love, speak, and shape their own lives.

Freedom does not mean doing anything without consequence. Freedom must be balanced with the freedom and safety of others. My freedom does not include the right to harm you. Your freedom does not include the right to enslave, abuse, threaten, or silence me.

But without freedom, morality becomes hollow.

A forced prayer is not faith. A forced marriage is not love. A forced confession is not truth. A forced obedience is not virtue. A forced silence is not peace.

A moral society must protect freedom of conscience, freedom of thought, freedom of speech, freedom from religious coercion, freedom of bodily autonomy, and freedom to live without humiliation.

God is not needed to defend these freedoms.

Human dignity is enough.

Seek Justice Without Revenge

Compassion without justice can become soft and ineffective. Justice without compassion can become cold and cruel.

A better morality needs both.

Justice begins with the recognition that harm must be answered. The victim matters. The truth matters. The wound matters. The abuse of power matters. The stolen freedom matters. The broken trust matters. The social conditions that made harm possible also matter.

Justice is not revenge. Revenge says, "You suffered, so now I want suffering in return." Justice says, "Harm was done, and the response must protect dignity, restore truth where possible, prevent future harm, and treat all persons fairly."

This distinction is important.

Religious systems often confuse justice with punishment. Hell is the most extreme example: infinite punishment for finite human wrongdoing or even honest disbelief. That is not justice. That is cruelty without proportion.

A humane morality must reject cruelty even when cruelty is aimed at the guilty.

This does not mean offenders should escape consequences. A society must protect people from violence, exploitation, fraud, abuse, and danger. Some people must be restrained. Some must be punished. Some must make restitution. Some need treatment. Some need long-term separation from society.

But punishment should have moral limits.

It should not become torture. It should not become humiliation for its own sake. It should not be driven by bloodlust. It should not be used to protect privilege. It should not punish the poor more harshly than the powerful. It should not excuse religious violence while condemning ordinary crime.

Justice must also be broader than courts and prisons.

There is injustice when a girl is denied education. There is injustice when a worker is exploited. There is injustice when caste decides dignity. There is injustice when women are controlled by religious patriarchy. There is injustice when children are indoctrinated with terror. There is injustice when blasphemy laws silence honest thought. There is injustice when medical care depends on wealth. There is injustice when the powerful use God to protect their power.

A better morality does not wait for heaven to correct injustice.

It demands justice here.

Reject Sacred Hierarchy

One of religion's recurring moral failures is the creation or protection of hierarchy.

Men over women. Believers over unbelievers. Priests over ordinary people. Upper caste over lower caste. One religion over another. Straight people over sexual minorities. Adults over children without respect for the child's inner life. The "chosen" over the "lost."

These hierarchies are often defended as natural, sacred, traditional, or divinely ordered.

A better morality rejects sacred hierarchy.

No person is born with greater dignity because of caste. No person is born closer to truth because of sex. No person deserves more freedom because of religion. No child deserves fear because adults call it faith. No woman deserves submission because a book says so. No unbeliever deserves hatred because he does not bow.

Equality does not mean that all people are identical. People differ in ability, character, wisdom, knowledge, responsibility, and action. A child and an adult do not have the same duties. A trained surgeon and an untrained person do not have the same role in an operating room. A criminal and a victim do not stand in the same moral position regarding a crime.

Equality means equal basic dignity, equal moral concern, equal protection under fair law, equal freedom from humiliation, equal right to conscience, and equal claim to be treated as a person.

This principle is impossible to maintain if sacred authority is allowed to rank human worth.

A secular morality can say clearly: no hierarchy is holy merely because it is old.

Every hierarchy must justify itself by human consequences.

If it humiliates, oppresses, silences, or exploits, it must fall.

A Morality That Can Admit Mistakes

One of the great strengths of secular morality is that it can learn from evidence.

If a policy increases suffering, we can study it. If a punishment fails to reduce crime, we can question it. If a teaching harms children, we can measure the harm. If discrimination damages mental health, education, income, or safety, we can see it. If a medical practice works, we can adopt it. If it fails, we can abandon it. If a tradition claims to protect society but produces fear, violence, or inequality, we can examine the evidence.

This matters because human beings are often wrong.

We are wrong about consequences, other groups, children, women, sexuality, mental illness, crime, addiction, punishment, and what helps people flourish.

A moral system that cannot admit error is dangerous.

Religious morality often has difficulty correcting itself because its rules are tied to sacred authority. If a command is said to come from God, changing it can feel like rebellion. If a scripture is considered perfect, questioning it can feel sinful. If a tradition is called holy, reform can feel like betrayal.

This slows moral progress.

A secular morality can say: we were wrong.

We were wrong about slavery. We were wrong about women. We were wrong about caste. We were wrong about children. We were wrong about mental illness. We were wrong about sexuality. We were wrong about punishment. We were wrong about animals. We were wrong about the environment.

This admission is not weakness.

It is the beginning of improvement.

A morality that can be corrected is better than a morality that calls its mistakes eternal.

Human Well-Being Is More Than Pleasure

A better morality should care about human well-being.

But well-being does not mean shallow pleasure. It does not mean constant happiness. It does not mean comfort at any cost.

A good life includes many things: health, safety, love, friendship, knowledge, freedom, dignity, purpose, creativity, rest, truth, justice, and the chance to develop one's abilities.

Sometimes well-being requires sacrifice. A parent loses sleep for a child. A doctor tells a painful truth to protect a patient. A citizen resists injustice at personal cost. A student struggles through difficulty to gain skill. A reformer faces danger to protect future generations.

So a morality based on well-being is not selfish ease. It is concern for the conditions that allow conscious beings to live with less suffering and more dignity.

This kind of morality asks practical questions. Can this child learn? Can this woman live safely? Can this worker participate with dignity? Can this disabled person flourish? Can this patient receive care? Can this citizen speak freely? Can this person leave a religion without fear? Can this human life open, grow, and breathe?

These questions are deeply moral. They are also practical.

They force us to look at real conditions, not just slogans.

A society may claim people are free, but if women are uneducated, the poor are trapped, dissenters are threatened, and children are indoctrinated by fear, freedom is incomplete.

A better morality asks not only what people are allowed to do in theory, but what they are actually able to become in life.

The Moral Test of Any Belief

Every belief should face a moral test.

Does it make people kinder? Does it make people more honest? Does it reduce suffering? Does it protect children? Does it respect women? Does it reject caste and sacred hierarchy? Does it allow doubt? Does it defend freedom of conscience? Does it treat unbelievers as full human beings? Does it respect evidence? Does it allow correction? Does it protect the weak from the powerful? Does it help people flourish in this life?

If a belief fails these tests, calling it religious does not save it.

If a belief frightens children, it is harmful. If a belief humiliates women, it is unjust. If a belief protects caste, it is immoral. If a belief threatens doubters, it is authoritarian. If a belief punishes blasphemy, it is afraid of thought. If a belief demands obedience over compassion, it is dangerous.

A better morality must be willing to say no.

No to sacred cruelty. No to holy hierarchy. No to divine excuses. No to fear-based obedience. No to the idea that morality belongs to God rather than to the conscious beings who suffer under moral decisions.

This is not hatred of religion.

It is loyalty to humanity.

The World Religion Failed to Build

If divine morality were truly superior, religious history should look very different.

The most religious societies should have been the most just. The most scripture-guided cultures should have been the most equal. The most priestly authorities should have protected the weak most consistently. The fear of God should have ended cruelty.

It did not.

Religion built temples, rituals, scriptures, monasteries, songs, art, charity, and communities.

But it also built walls. It built caste. It built patriarchy. It built blasphemy laws. It built fear of hell. It built holy war. It built persecution. It built obedience to authority. It built shame around the body. It built suspicion of outsiders. It built systems where children were taught what to believe before they were taught how to think.

This mixed record shows that religion is not the moral foundation it claims to be.

Human beings must build something better.

Not perfect.

Better.

A morality that begins with real suffering. A morality that protects dignity. A morality that values freedom. A morality that seeks justice. A morality that listens to evidence. A morality that can say, “We were wrong.” A morality that does not need God to care for the world.

What a Godless Morality Asks of Us

A godless morality is not easier.

In some ways, it is harder.

It does not allow us to hand responsibility to heaven. It does not allow us to say, “God will judge.” It does not allow us to wait for the next life to correct this one. It does not allow us to hide cruelty behind scripture. It does not allow us to excuse injustice as divine plan. It does not allow us to say, “This suffering must have some heavenly purpose.”

Instead, it asks us to act.

If children are hungry, feed them. If people are sick, treat them. If women are oppressed, defend equality. If caste humiliates people, reject caste. If religion frightens children, protect children. If laws are unjust, change them. If prisons are cruel, reform them. If poverty crushes dignity, reduce poverty. If lies spread, defend truth. If science saves lives, support science.

If no god will do these things for us, then we must do them for one another.

That is not moral emptiness.

That is moral seriousness.

A world without God is not a world without duty.

It is a world where duty belongs fully to us.

The Beauty of Human Responsibility

There is beauty in this view.

No angel may come. No heaven may intervene. No divine hand may stop the bullet, cure the disease, feed the hungry, or protect the child.

But human hands can do some of this.

Human minds can. Human courage can. Human compassion can. Human science can. Human law can. Human solidarity can.

A nurse can comfort. A doctor can heal. A teacher can open a mind. A judge can protect rights. A citizen can resist injustice. A parent can break a cycle of fear. A writer can challenge cruelty. A scientist can discover a cure. A stranger can stop and help.

These acts do not become smaller because God is absent.

They become larger.

They show that human beings can create meaning, justice, and care without divine command.

A godless morality does not lower human duty.

It raises it.

It does not say life is empty. It says life is ours. It says this world matters because this is where suffering happens and where suffering can be reduced. It says human dignity matters because conscious life is precious. It says justice matters because victims cannot wait for heaven. It says freedom matters because each person has only one life that we know of. It says compassion matters because no one is coming from the sky to do our work for us.

That is a serious morality.

That is a beautiful morality.

That is morality without God.

Chapter Summary

This chapter presented a positive alternative to divine-command morality.

A better morality begins with real suffering, human dignity, freedom, justice, equality, evidence, and well-being. It does not ask first what God commands. It asks who is harmed, who is helped, whose dignity is protected, whose freedom is denied, and what kind of life conscious beings are able to live.

Human dignity means that every person has worth before religion, caste, gender, nationality, belief, status, or usefulness. Freedom matters because human lives are lived from the inside. Justice matters because harm must be answered without becoming revenge. Equality matters because no sacred hierarchy should rank human worth.

A secular morality can learn from evidence and correct itself. It can admit error. It can change when old beliefs cause harm. It can place human well-being at the center without reducing life to shallow pleasure.

This morality is not perfect because human beings are not perfect. But it is more honest than divine command because it accepts responsibility. It does not wait for heaven to fix the world. It asks human beings to reduce suffering and protect dignity here and now.

Key Takeaways

A better morality begins with real suffering, not divine command.

Harm is not a matter of theology. Pain, fear, humiliation, and oppression matter because conscious beings experience them.

Human dignity does not depend on religion, caste, gender, belief, status, or usefulness.

Freedom matters because each person's life is lived from the inside.

Justice is not revenge. It protects dignity, truth, fairness, and future safety.

Secular morality can improve because it can admit mistakes and learn from evidence.

Human well-being includes health, safety, love, dignity, freedom, justice, truth, and the ability to flourish.

A godless morality is not moral emptiness. It is human responsibility.

Transition to the Next Chapter

If morality does not require God, another fear remains.

Many people believe that without heaven, life loses meaning. If death is final, they ask, what is the point? If there is no eternal soul, why should love, work, kindness, sacrifice, and beauty matter?

This fear is understandable.

But it is mistaken.

A flower is not worthless because it fades. A song is not meaningless because it ends. A human life is not empty because it is brief.

The next chapter will examine meaning without heaven, and why a mortal life can still be deep, beautiful, and worth living.

Sources

Peter Singer, *Practical Ethics.*

Martha Nussbaum, *Creating Capabilities.*

Amartya Sen, *The Idea of Justice.*

John Rawls, *A Theory of Justice.*

A.C. Grayling, *The God Argument.*

Paul Kurtz, *Forbidden Fruit: The Ethics of Humanism.*

Stanford Encyclopedia of Philosophy, "Dignity."

Stanford Encyclopedia of Philosophy, "Human Rights."

Stanford Encyclopedia of Philosophy, "Well-Being."

Internet Encyclopedia of Philosophy, "Sen's Capability Approach."

United Nations, *Universal Declaration of Human Rights.*

Chapter 11: Meaning Without Heaven

Why a Mortal Life Can Still Be Deep, Beautiful, and Worth Living

A flower is not worthless because it fades. A song is not meaningless because it ends. A sunset is not empty because darkness follows. A child's laughter is not trivial because childhood passes.

A human life is not meaningless because it is brief.

The fact that something ends does not prove that it never mattered.

Yet this is one of religion's strongest emotional arguments. Religion tells us that without heaven, life has no final meaning. Without an immortal soul, love is temporary. Without eternal reward, sacrifice is wasted. Without reunion after death, grief is unbearable. Without God, life is only a short biological accident between two darknesses.

This fear is powerful because it touches the deepest part of us.

We do not merely know that we will die. We feel the terror of it. We know that the people we love will die. We know that our own memories will someday disappear. We know that our bodies will age, weaken, and stop. We know that the house of the self, built slowly through childhood, work, love, struggle, and memory, will not stand forever.

Religion enters this fear and offers a promise: you will not truly die, your loved ones are waiting, the soul will continue, heaven will repair every wound, justice will be completed, and love will become eternal.

For many people, this is the most emotionally difficult religious belief to question. One can doubt a scripture, reject a miracle, accept evolution, and still feel afraid to let go of heaven.

Because heaven is not only an idea.

It is a shelter built over grief.

This chapter will not mock that shelter. Grief deserves tenderness. Death deserves honesty. The longing to see the dead again is one of the most human feelings we have.

But longing is not evidence.

A belief may comfort us and still be false.

The question is not whether heaven feels consoling. The question is whether life loses meaning without it.

And the answer is no.

A mortal life can still be deep, beautiful, serious, loving, and worth living.

Meaning does not require eternity. Meaning requires value.

And value can exist in time.

The Fear That Death Makes Life Empty

Death is not a small problem.

It is the great boundary of human life.

Everything we do happens under its shadow. We build families, learn skills, write books, heal patients, raise children, plant trees, fight injustice, forgive enemies, and hold loved ones, knowing that all of it is temporary.

No wonder human beings invented afterlives.

The mind struggles to imagine its own absence. We can imagine a dark room. We can imagine silence. We can imagine being alone. But we cannot truly imagine nonexistence, because imagination itself requires a mind that exists.

So the mind resists. It says, “I cannot simply end.” It says, “Something of me must continue.” It says, “The people I love cannot be gone forever.”

Religion gives shape to this resistance. It turns fear into doctrine and longing into geography. Heaven becomes a country of repaired loss. The soul becomes the traveler. Death becomes a doorway.

This is emotionally understandable. But emotional need is not proof.

A thirsty person may dream of water in a desert. The dream is understandable, but it does not fill the canteen. A grieving person may dream of reunion. The dream may comfort, but it does not prove another world. A frightened person may long for immortality. The longing is human, but it does not make the soul real.

The fact that death is frightening does not prove heaven. The fact that grief is painful does not prove reunion. The fact that life is brief does not prove it is meaningless.

It proves only that life is fragile.

And fragile things can still be precious.

Why Temporary Things Can Matter Deeply

Much of what we value is temporary.

A meal ends. A concert ends. A childhood ends. A friendship may end. A marriage may end through death.

A vacation ends. A day with family ends. A book ends. A life ends.

But we do not normally say these things are meaningless because they end.

A beautiful song is not ruined by the final note. The final note is part of what gives the song shape. A good meal is not meaningless because the plate is eventually empty. A child's birthday is not pointless because the child will grow older. A conversation with a dying patient is not worthless because the patient may die tomorrow. A mother holding her newborn is not meaningless because both mother and child are mortal.

Meaning does not depend on endless duration.

In fact, endless duration can sometimes weaken meaning. If every moment were repeated forever, urgency might disappear. If every mistake could be repaired in eternity, responsibility might feel lighter. If every goodbye were temporary, farewell would lose some of its painful truth. If every life continued forever, the rare and delicate beauty of one lifetime might be harder to feel.

Finitude gives shape to life.

A poem matters partly because it has lines, limits, rhythm, and ending. A painting matters partly because the frame gives it form. A human life matters partly because time is limited.

We do not need eternity for meaning. We need depth. We need love. We need purpose. We need connection. We need truth. We need the chance to reduce suffering and leave something better than we found it.

A finite life can contain all of this.

Love Does Not Need Eternity

Religion often says that if love ends at death, then love is meaningless.

But this misunderstands love.

Love is not meaningful only if it lasts forever. Love is meaningful because it is experienced, given, received, remembered, and lived.

A mother who cares for a child for one day has done something meaningful. A husband who holds his dying wife's hand has done something meaningful. A friend who sits beside another friend during grief has done something meaningful. A doctor who relieves pain in a patient who will die soon has done something meaningful.

None of these acts require eternity.

Their value is in the care itself.

Imagine a child in pain. You comfort the child. The child later stops crying. That moment of comfort mattered. It does not become meaningless because the moment passed. It does not become meaningless because the child will someday die. It mattered because suffering was reduced, fear was softened, and love was given.

Love is not a contract with eternity. Love is a response to the living.

This is important because religion often holds love hostage. It says love must be eternal to be real. It says loved ones must survive death somewhere, or else love is defeated.

But love is not defeated by mortality. Love is made more urgent by mortality.

Because life is brief, we must speak kindly now. Because people die, we must not postpone affection. Because memory fades, we must create tenderness while we can. Because there may be no heaven, the way we treat one another here becomes even more important.

If this is the only life we know we have, then love is not less valuable.

It is more valuable.

The Empty Chair

Grief is where atheism is tested most deeply.

It is one thing to discuss mortality in philosophy. It is another to stand beside a hospital bed, a funeral pyre, a grave, or an empty room.

A person dies, and the world becomes strangely wrong.

The toothbrush remains. The clothes remain. The photographs remain. The chair remains. The favorite cup remains in the kitchen.

The voice is gone.

The body may be gone, but habits continue searching. The hand reaches for a phone to call someone who will not answer. The mind hears a sentence and thinks, "I must tell him," and then remembers.

Religion offers a simple comfort: the person is not really gone.

Atheism cannot honestly offer that.

It must say something harder: the person has died. The living presence has ended. There is no reliable evidence that an immortal soul continues elsewhere.

That truth hurts.

But grief without illusion is not loveless. It may even be more faithful to reality.

The dead live in the effects they had on the world. They live in the love they gave, the children they raised, the students they taught, the patients they healed, the kindness they showed, the ideas they shared, the habits they shaped, the memories they left, and the changes they made in other human beings.

This is not literal immortality. It is influence. It is legacy. It is memory. It is the way one life continues as a cause inside other lives.

When we remember the dead honestly, we do not need to pretend they are watching from another realm. We honor them by carrying forward what was good, learning from what was painful, and living more fully because they once lived.

Grief does not require heaven to be sacred in the human sense.

Love makes grief sacred. Memory makes grief sacred. The reality of loss makes grief sacred.

No supernatural world is needed.

Legacy Without Immortality

People often think there are only two choices: eternal life or nothing.

But that is too narrow.

A person does not need to live forever to matter after death.

Every life leaves traces.

Some traces are large and visible: books, discoveries, inventions, institutions, movements, children, public service, art, medicine, science, law, reform.

Most traces are quieter: a word of encouragement that changed a student's confidence, a kindness that helped a friend survive despair, a habit of honesty that shaped a child, a patient comforted during fear, a family protected by sacrifice, a prejudice challenged, a question planted, a wound healed, a meal served, a hand held.

These are not eternal in the religious sense. But they are real.

Human beings live inside chains of influence. We are shaped by people who came before us, and we shape people who come after us. No one is completely self-created. Our language, knowledge, values, habits, stories, and opportunities come from countless lives, most of them forgotten by name.

The farmer who grew food, the teacher who taught letters, the scientist who discovered a vaccine, the parent who sacrificed, the worker who built roads, the nurse who stayed awake at night, the reformer who challenged injustice, and the writer who preserved truth all continue through effects.

This is symbolic immortality, not literal immortality. It is not the soul surviving. It is the life echoing.

That desire is understandable. We want our lives to matter beyond our last breath. We want the good we did to continue. We want love to leave a mark.

That desire does not prove heaven.

It shows that human beings need meaning, connection, and continuity.

We can create those without believing in an immortal soul.

A life can matter because of what it gives to other lives.

This Life Is Not a Waiting Room

One of religion's most harmful ideas is that this life is only a preparation for another life.

A test. A waiting room. A temporary station. A place of trial before the real destination.

This idea may comfort people who suffer. It may help them endure poverty, injustice, illness, grief, or oppression. But it can also be dangerous.

If this life is only a test, present suffering may be minimized. If heaven will fix everything later, injustice here may feel less urgent. If the soul matters more than the body, bodily pain may be neglected. If obedience earns eternal reward, freedom may be sacrificed. If the next world is more important than this one, people may tolerate cruelty in this one.

A child frightened by hell may be told it is for eternal salvation. A woman denied freedom may be told her reward is in heaven. A poor man may be told to accept suffering because God will reward him later. A patient may be told to pray when medical care is needed. A society may tolerate injustice because divine justice is expected after death.

This is not harmless.

A life treated as a waiting room can be wasted.

A better view says this life is not a rehearsal.

This is the life in which children need food. This is the life in which patients need care. This is the life in which women need freedom. This is the life in which caste must be rejected. This is the life in which truth must be defended. This is the life in which love must be expressed. This is the life in which suffering must be reduced. This is the life in which justice must be done.

If another world is not guaranteed, then this world becomes more important, not less.

Atheism does not make life meaningless.

It makes life immediate.

Meaning Is Made, Not Given

Religion often says that meaning must be given from above. God gives purpose. Scripture gives direction. Heaven gives destination. Divine plan gives structure. Without this, religion says, human beings are lost.

But meaning does not have to be given from outside life.

Meaning can be made within life.

A physician finds meaning in healing. A teacher finds meaning in opening minds. A parent finds meaning in raising children. A scientist finds meaning in discovering truth. A reformer finds meaning in fighting injustice. An artist finds meaning in beauty. A friend finds meaning in loyalty. A humanist finds meaning in reducing suffering. A writer finds meaning in leaving words that may help strangers think more clearly.

These meanings are not fake because they are human-made.

A house is human-made. It can still shelter a family. A medicine is human-made. It can still save a life. A song is human-made. It can still move the heart. A constitution is human-made. It can still protect freedom. A moral purpose is human-made. It can still be noble.

The idea that meaning must come from God assumes that human beings cannot create anything truly valuable.

That is false.

Human beings create language, music, medicine, mathematics, friendship, justice, science, literature, and love-filled homes. These are not worthless because they are human.

They are precious because they are human.

Meaning is not something lowered from heaven.

Meaning is something built from life.

The Myth That Eternity Solves Meaning

Many people assume that eternal life would automatically make life meaningful.

But would it?

If a life is meaningless for seventy years, why would it become meaningful by continuing forever?

Endless time does not create purpose by itself. A pointless activity repeated eternally does not become meaningful. An empty life stretched forever is still empty. Immortality may remove the fear of ending, but it does not automatically answer what life is for.

In fact, immortality creates its own problems.

Would choices matter as much if time never ended? Would urgency survive if every delay could be corrected later? Would courage mean the same thing if nothing final could be lost? Would love have the same tenderness if parting were never real? Would life become heavy, repetitive, or exhausted after endless time?

The key point is simple.

Eternity alone does not give meaning.

Meaning comes from what is valued: love, truth, beauty, justice, compassion, growth, understanding, courage, and contribution.

These can exist in a finite life.

Therefore, heaven is not necessary for meaning.

Science, Wonder, and Meaning

Some people fear that science makes life cold. They think that if we are products of evolution, if the mind comes from the brain, if morality comes from human life, and if death is final, then wonder disappears.

But science does not erase wonder. It enlarges it.

The atoms in our bodies were forged in stars. Life on Earth evolved over billions of years. The human brain contains billions of neurons and trillions of connections. A child learns language from sound and love. A wound heals through cellular cooperation. A tree turns sunlight into living tissue. A bird crosses continents. A patient survives because generations of scientists, physicians, nurses, and researchers built knowledge piece by piece.

This is not a meaningless universe. It is a universe that did not design us, but still gave rise to beings who can ask about meaning.

We are not the center of creation. But we are centers of experience.

We are small in the cosmos, but not small to one another.

To a child, a parent's love is not small. To a patient, relief from pain is not small. To a grieving person, one act of kindness is not small. To a victim, justice is not small. To a lonely person, friendship is not small.

Cosmic size is not the measure of meaning.

Human experience is.

We do not need to be the purpose of the universe for our lives to have purpose.

We can create purpose inside the universe.

Mortality Makes Life More Precious

Mortality is painful.

But it also sharpens value.

Because time is limited, a day matters. Because childhood passes, attention matters. Because parents age, phone calls matter. Because health can change, gratitude matters. Because memory can fade, presence matters. Because death is real, love should not be postponed.

A person who believes life is eternal may sometimes delay too much.

I will forgive later. I will speak later. I will change later. I will live fully later. I will question later. I will love later.

But later is not guaranteed.

Mortality teaches urgency.

It says: do not waste your only known life in fear. Do not surrender your mind to inherited dogma. Do not let religion make you ashamed of being human. Do not wait for heaven to give permission to value Earth. Do not postpone kindness. Do not postpone truth. Do not postpone freedom. Do not postpone love.

This is not despair. It is awakening.

The shortness of life does not make it empty. It makes it more precious.

A rare thing is not worthless because it is rare. A brief thing is not meaningless because it is brief. A fragile thing is not insignificant because it can break.

Life matters because it is lived.

And because it is lived only once, as far as we know, it matters intensely.

Meaning After Leaving Religion

For many people, leaving religion is not only an intellectual change. It is an emotional earthquake.

The old map disappears. The old purpose disappears. The old fear may remain. The old community may reject them. The promise of heaven fades. The fear of hell may still echo.

At first, atheism may feel empty.

This is understandable.

If a person has been told since childhood that God is the source of meaning, then life without God may feel like a room after the lights go out.

But slowly, another kind of light appears. Not from heaven. From reality.

The morning is still beautiful. Food still tastes good. Children still laugh. Books still open minds. Music still moves the heart. The sick still need care. The grieving still need comfort. The oppressed still need freedom. Friends still need loyalty. The mind still wants truth. The body still needs kindness. The world still contains work worth doing.

Leaving religion does not remove meaning. It removes an imposed meaning. Then a person must build a more honest one.

This can be frightening. But it can also be freeing.

Instead of being told that life is a test, one can see life as a chance: a chance to understand, to love, to heal, to create, to reduce suffering, to leave the world slightly better, and to live without fear of imaginary punishment.

Science-based atheism does not say, "There is no meaning."

It says, "Meaning is our responsibility."

That is harder than receiving meaning from authority.

But it is also more mature.

The Meaning of Helping

If a person asks, "What is the meaning of life?" the question can feel too large.

But if a person asks, "Can I reduce suffering somewhere?" the path becomes clearer.

A hungry child needs food. A lonely person needs company. A patient needs relief. A student needs encouragement. A victim needs justice. A frightened mind needs truth. A grieving friend needs presence. A society needs courage.

Meaning often begins in service to something beyond the narrow self.

This does not require religion. It requires attention to reality.

A doctor treating pneumonia does not need heaven to make the treatment meaningful. A scientist developing a vaccine does not need God to make the discovery meaningful. A teacher helping a poor child learn to read does not need scripture to make the work meaningful. A reformer fighting caste discrimination does not need divine command to make justice meaningful. A parent comforting a crying child does not need immortality to make the love meaningful.

Helping matters because suffering matters.

This is one of the clearest foundations of meaning.

When we reduce suffering, life becomes more meaningful, not because eternity rewards us, but because a real burden in a real life becomes lighter.

That is enough.

Meaning Through Truth

Truth can be painful. It can take away comforting beliefs. It can remove illusions. It can force us to face mortality, uncertainty, and responsibility.

But truth also gives dignity.

A life built on falsehood may feel safe, but it rests on weak ground. A life built on truth may feel harder, but it is more honest.

There is meaning in refusing to deceive oneself. There is meaning in saying, "I will not believe simply because I am afraid." There is meaning in saying, "I will not call a story true merely because it comforts me." There is meaning in saying, "I will not teach a child fear as fact." There is meaning in saying, "I will face death honestly and still love life."

This is one of the deepest forms of courage.

Religion often says faith is courage. Sometimes faith may require courage, especially when it motivates compassion or resistance to injustice. But believing what comforts us is not always courage.

Sometimes courage means letting go of comfort. Sometimes courage means standing without heaven. Sometimes courage means admitting that the dead are gone and still honoring them. Sometimes courage means accepting that life is finite and still living generously. Sometimes courage means rejecting false hope and choosing real love.

Truth is not cold.

Truth can be a form of respect for life.

The Heaven We Make for One Another

If there is no heaven, then the word "heaven" can still teach us something.

It points to what human beings long for: peace, safety, love, reunion, justice, rest, freedom from pain, freedom from fear, a place where tears are wiped away.

Religion places that hope in another world.

Humanism asks us to bring as much of it as possible into this one.

Feed the hungry. Comfort the grieving. Protect the child. Free the oppressed. Care for the sick. Welcome the lonely. Defend the doubter. Reduce fear. Increase justice. Create beauty. Tell the truth. Love while we can.

We may not be able to create heaven. But we can create moments of shelter.

A hospital room where pain is relieved. A home where a child is safe. A classroom where a girl is encouraged. A court where the weak are protected. A friendship where grief is held. A society where people need not fear religion. A world where dignity does not depend on belief.

These are not eternal heavens.

They are human heavens.

Brief, imperfect, fragile, and real.

That is where meaning lives.

Not above the clouds.

Among us.

Chapter Summary

This chapter examined meaning without heaven.

Religion often claims that life is meaningless without an immortal soul, an afterlife, or eternal reward. This fear is understandable because death, grief, and finality are emotionally difficult. But the fact that life ends does not mean life is empty.

Many temporary things are meaningful: songs, meals, childhood, friendship, love, healing, kindness, and courage. Their value does not depend on lasting forever.

Love does not need eternity to be real. Grief does not need heaven to be sacred in the human sense. Legacy does not require an immortal soul. A life can continue through influence, memory, kindness, children, ideas, healing, and the difference it makes in other lives.

This life is not a waiting room for another world. It is the only life we know we have. That makes truth, justice, love, freedom, and compassion more urgent, not less.

Meaning is not handed down from heaven. It is built through love, work, service, truth, beauty, courage, and the reduction of suffering.

A mortal life can still be deep, beautiful, and worth living.

Key Takeaways

Death is frightening, but fear of death does not prove heaven.

Temporary things can be deeply meaningful.

Love is not meaningless because it ends at death.

Grief without belief in heaven can still be tender, honest, and sacred in a human sense.

Legacy does not require immortality. A life can continue through its effects on others.

This life is not a rehearsal or waiting room. It is where suffering must be reduced and love must be expressed.

Meaning is made through values, relationships, truth, service, justice, and care.

Mortality does not make life empty. It makes life precious.

Transition to the Next Chapter

We have now examined the three great fears religion uses to defend itself.

Without a soul, religion says, the self disappears. Without God, religion says, morality collapses. Without heaven, religion says, life becomes meaningless.

But the evidence points in another direction.

The brain creates the self. Human life creates morality. This world gives us enough reason to love, care, think, help, and build meaning.

The next chapter will bring these arguments together.

It will ask what remains after God, soul, and heaven are no longer needed.

The answer is not emptiness.

The answer is the human being: natural, mortal, fragile, responsible, and still deeply worthy.

Sources

Irvin D. Yalom, *Staring at the Sun: Overcoming the Terror of Death.*

Viktor E. Frankl, *Man's Search for Meaning.*

Albert Camus, *The Myth of Sisyphus.*

Bertrand Russell, "A Free Man's Worship."

Carl Sagan, *The Demon-Haunted World.*

Thaddeus Metz, "The Meaning of Life," *Stanford Encyclopedia of Philosophy.*

Steven Luper, "Death," *Stanford Encyclopedia of Philosophy.*

I. J. Peltomäki, "Philosophy of Meaning in Life in the Light of Finitude," 2023.

M. Svet et al., "Applying Terror Management Theory to Patients With Life-Threatening Illnesses," 2023.

Ernest Becker, *The Denial of Death.*

Chapter 12: The Human Being After God

Mind Without Soul, Morality Without God, Meaning Without Immortality

When the soul disappears, the human being does not disappear.

The child still laughs. The mother still loves. The patient still suffers. The victim still deserves justice. The stranger still deserves compassion. The dying person still deserves tenderness. The mind still wonders. The heart still breaks. The hand still reaches for another hand in the dark.

What disappears is not humanity.

What disappears is the illusion that humanity needs supernatural permission to matter.

For thousands of years, religion told human beings that they were valuable because God made them, that they were moral because God commanded them, and that life mattered because the soul would continue after death. It placed human dignity outside the world, in a divine creator, a spiritual essence, and an eternal future.

This sounded comforting. But it also made human value dependent on beliefs that have never been proven. It made morality dependent on ancient commands. It made meaning dependent on another world.

This book has argued for a different view.

The self does not need a soul. It arises from the living brain, body, memory, emotion, language, relationship, and time.

Morality does not need God. It arises from care, cooperation, empathy, fairness, suffering, dignity, reason, and the need to live together.

Meaning does not need heaven. It arises from love, truth, work, beauty, service, justice, memory, and the difference one life can make in other lives.

This is not a fall from dignity. It is dignity brought home.

It places human value not in an invisible realm, but in the real world where people are born, loved, wounded, healed, educated, oppressed, freed, remembered, and missed.

A human being after God is not empty.

A human being after God is natural, mortal, fragile, responsible, and still deeply worthy.

We Are Natural, But Not Empty

Religion often teaches that if human beings are natural, they are "only" matter: only flesh, only chemistry, only neurons, only animals, only dust.

That little word "only" carries a great insult. It suggests that nature is low, that matter is dead, that biology is cheap, and that human life needs something supernatural before it can have value.

But why should that be true?

A forest is natural, but it is not empty. A newborn child is natural, but not worthless. A rose is natural, but not ugly. A brain is natural, but it can create mathematics, music, memory, medicine, poetry, grief, justice, and love.

The mistake is not in saying we are natural. The mistake is in thinking nature is small.

The universe has produced stars, oceans, forests, cells, nervous systems, eyes, language, laughter, moral protest, and minds capable of studying the universe itself. We are not separate from nature. We are nature becoming conscious in one fragile form.

That is not an insult.

That is astonishing.

A human being is not made less precious because the brain creates the mind. A song is not less beautiful because it is made of sound waves. A rainbow is not less beautiful because light can be explained. A flower is not less beautiful because it grows from soil, genes, sunlight, and water.

Explanation does not destroy wonder. It deepens it.

The soul was an ancient attempt to protect human wonder. But the soul is not needed. The brain itself is wondrous enough.

The self is not a ghost inside a machine. The self is a living process: body, brain, memory, feeling, and story moving through time. It can love, fear, imagine, change, suffer, learn, apologize, and begin again.

No supernatural essence is needed for any of this.

We Are Bodies, But Not Mere Objects

Religion often separates the body and the soul. The body is temporary, but the soul is eternal. The body is flesh, but the soul is spirit. The body is low, but the soul is high.

This division has caused enormous harm. It has taught people to mistrust the body, shame desire, fear pleasure, neglect health, and treat bodily life as less important than spiritual obedience.

But we are not souls trapped inside bodies. We are embodied beings.

The body is not a prison for the self. The body is part of the self.

Hunger affects mood. Pain affects patience. Sleep affects judgment. Hormones affect emotion. Illness affects identity. Touch can comfort. Aging can change the story of a life. A damaged brain can alter memory, personality, and awareness. A healthy body can support freedom, learning, love, and moral action.

This is why a science-based view of the human being is not cold. It is tender. It tells us to care for the body because the body is where life is lived.

Feed the hungry. Treat the sick. Protect children. Prevent injury. Relieve pain. Support mental health. Care for the aging. Respect disability. Help the addicted. Comfort the dying.

These are not secondary matters below spiritual salvation. They are central moral duties.

If the mind depends on the brain, then brain health matters. If consciousness depends on living conditions, then those conditions matter. If suffering happens in bodies, then bodies matter.

Religion has often told people to save souls.

Humanism tells us to save lives.

The first can become abstract. The second is immediate.

We Are Mortal, But Not Meaningless

We will die.

This is the truth religion most desperately tries to soften.

The brain that creates the self will one day stop. The memories will no longer be experienced. The voice will become silent. The body will return to the world. The person, as a living center of experience, will end.

This is painful.

But it does not make life meaningless.

A life does not need to last forever to matter. A child's smile matters though it lasts a moment. A friendship matters though it may end. A marriage matters though death may part it. A meal given to the hungry matters though hunger will return elsewhere. A patient comforted near death matters though the patient dies. A book matters though paper decays. A song matters though the last note fades.

Meaning is not measured by endless duration.

Meaning is measured by value.

A finite life can contain love, courage, kindness, discovery, forgiveness, creativity, and justice. These do not become worthless because they are temporary.

In fact, mortality can make them more precious.

Because time is limited, attention matters. Because people die, kindness matters. Because childhood passes, care matters. Because health can change, gratitude matters. Because memory can fade, presence matters.

Because there may be no heaven, this life matters more, not less.

We do not need to be eternal to have meaning. We need lives that touch value.

A mortal being who loves honestly, reduces suffering, seeks truth, protects dignity, and leaves even a small kindness behind has not lived meaninglessly.

Such a life may be brief.

But it is real.

And real meaning is better than imaginary eternity.

We Are Moral Because Others Can Suffer

Religion says morality needs God. But morality begins with a simpler and stronger fact: others can suffer.

A child can be frightened. A woman can be beaten. A prisoner can be tortured. A patient can be neglected. A poor family can be hungry. A lower caste person can be humiliated. A nonbeliever can be persecuted. A lonely person can despair. A dying person can need comfort.

These realities do not need divine approval before they matter.

They matter because suffering is real.

Once we understand that other beings have inner lives, morality begins. Your pain matters to you. My pain matters to me. A child's pain matters to the child. A stranger's fear matters to the stranger. A victim's humiliation matters even if society ignores it.

This is the moral foundation religion often hides under divine language.

Do not torture, because torture causes terrible suffering. Do not enslave, because slavery crushes freedom and dignity. Do not humiliate, because humiliation wounds the self. Do not indoctrinate children with terror, because fear can damage the developing mind. Do not persecute unbelievers, because conscience must be free. Do not oppress women, because women are full human beings. Do not defend caste, because birth does not determine dignity.

No God is required to understand this.

In fact, divine command can sometimes confuse it. If a scripture permits cruelty, a believer may excuse it. If a prophet commands violence, a follower may sanctify it. If tradition ranks people, a community may call inequality sacred.

A morality based on suffering and dignity has a great advantage: it can judge every command by its effect on real lives.

It can ask: who is harmed, who is silenced, who is afraid, who is humiliated, who benefits, and who pays the price?

These questions are better than asking whether a rule is ancient. They are better than asking whether a priest approves. They are better than asking whether a holy book protects the powerful.

Moral life does not need a supernatural judge.

It needs human beings willing to see suffering clearly and respond with courage.

We Are Responsible Because No God Will Rescue Us

A godless world is not a world without responsibility.

It is a world where responsibility can no longer be transferred upward.

If there is no divine rescuer, then human beings must rescue one another. If there is no heavenly justice waiting to repair the world, then justice must be pursued here. If there is no god stopping cruelty, then we must stop it. If there is no divine plan behind suffering, then we must reduce suffering ourselves.

If no supernatural parent is watching over humanity, then humanity must grow up.

This is difficult.

Religion often gives people the feeling that someone is ultimately in control. Even when life is painful, believers may say, "God has a plan." Even when injustice wins, they may say, "God will judge." Even when children suffer, they may say, "There is a reason."

Atheism removes that comfort.

But it also removes a dangerous excuse.

If suffering is not part of a divine plan, then suffering should not be romanticized. If injustice will not be corrected by heaven, then injustice must be confronted now. If prayer alone will not cure disease, then medicine, research, public health, and care matter. If God will not protect children from abuse, then families, schools, laws, and communities must protect them. If God will not end caste, patriarchy, racism, poverty, war, or religious hatred, then human courage must.

A godless world is not morally empty.

It is morally urgent.

This urgency is one of the strengths of science-based atheism. It says: stop waiting for supernatural rescue. Feed the hungry. Teach the child. Treat the patient. Comfort the grieving. Question the priest. Challenge the tyrant. Reject the hierarchy. Defend the dissenter. Protect the unbeliever. Reduce suffering while there is still time.

That is not despair.

That is responsibility.

The Final Humanist Answer

Humanism begins with the human being, not as a perfect creature, but as a vulnerable one.

A human being can reason, but also deceive himself. A human being can love, but also hate. A human being can cooperate, but also dominate. A human being can create music, medicine, and justice, but also war, superstition, and cruelty.

Humanism does not worship humanity.

It takes responsibility for humanity.

That is an important difference.

Religion often says that without God, human beings become arrogant. But the opposite can be true. Science-based atheism can make us humbler. It reminds us that we are animals, products of evolution, dependent on the brain, shaped by culture, vulnerable to bias, and limited in knowledge.

It removes the fantasy that we are the center of creation.

But it does not remove value.

A human being can be small in the universe and still enormous to another human being. To a child, a loving parent is not small. To a patient, a caring physician is not small. To a lonely person, one faithful friend is not small. To a victim, justice is not small. To a dying person, a hand held gently is not small.

The universe may not have designed us, but we can design kinder societies. The universe may not judge us, but we can judge cruelty. The universe may not love us, but we can love one another. The universe may not give meaning, but we can create meaning.

The final humanist answer is not, "Human beings are gods."

The final humanist answer is, “Human beings are responsible.”

Responsible for truth, justice, compassion, and the fragile minds and bodies through which life is experienced. Responsible for this world because no other world has been shown to exist.

The Soul Was Too Small for the Human Being

The soul was meant to make human beings larger.

In many ways, it made them smaller.

It reduced the self to a hidden spiritual object. It made the body secondary. It made this life less important than another life. It made obedience more important than inquiry. It made morality dependent on command. It made death a doorway in imagination, but often made life a waiting room.

The real human being is richer than the soul.

The human being is not a little ghost riding inside the skull. The human being is a body that feels, a brain that remembers, a face that expresses, hands that work, a voice that speaks, a history that shapes, a language that carries meaning, a community that wounds or heals, and a consciousness that briefly lights a corner of the universe.

The soul is too simple for this.

It is too thin.

It cannot explain dementia. It cannot explain brain injury. It cannot explain sleep, anesthesia, and coma. It cannot explain how childhood shapes adulthood. It cannot explain why hunger changes mood or why pain changes patience. It cannot explain why memory builds identity. It cannot explain why love is embodied.

The brain-based view does not make the person smaller.

It makes the person more real.

It places the human being where the human being actually lives: in nature, in the body, in time, in relationship, in society, in memory, and in the world.

That is not a loss.

It is a return to reality.

God Was Too Unreliable for Morality

God was supposed to make morality secure.

But divine morality has never been secure.

Different religions claim different divine commands. Different scriptures contain different rules. Different believers interpret the same text in opposite ways. Religious authorities often change their moral teachings when society changes. Holy books have been used to defend both cruelty and compassion, both slavery and liberation, both patriarchy and equality, both peace and violence.

This is not a stable foundation.

It is a battlefield of interpretation.

The real moral work is always done by human judgment.

People must decide which commands to follow, which to ignore, which to reinterpret, and which to reject. They must decide whether a scripture's command is humane or cruel. They must decide whether tradition deserves respect or resistance.

Even the believer must use human conscience.

That means God is not doing the moral work.

Human beings are.

A better morality admits this openly. It does not pretend that an invisible authority has solved all moral questions. It does not hide behind sacred language. It does not excuse cruelty by calling it obedience. It does not make doubt a sin.

It asks hard human questions: what reduces suffering, what protects dignity, what respects freedom, what is fair, what can evidence show, who is being harmed, who is being silenced, who is being left out, and what must change?

This morality is not perfect, but it can improve.

A morality that can improve is better than a morality that calls ancient error eternal.

Heaven Was Too Distant for Meaning

Heaven was supposed to give life meaning.

But heaven also moved meaning away from Earth. It told people that the real home is elsewhere, the real justice is later, the real reward is after death, and the real life begins beyond the grave.

This can comfort suffering people, but it can also weaken responsibility.

If heaven will fix everything, why fight so hard here? If suffering has a divine purpose, why question it? If the soul is more important than the body, why care so much about earthly conditions? If eternity matters most, why defend temporary freedom?

A humanist view brings meaning back to Earth.

This is where children need protection. This is where patients need care. This is where women need equality. This is where caste must be rejected. This is where hunger must be fed. This is where truth must be defended. This is where love must be spoken. This is where books are written, wounds are healed, ideas are tested, songs are sung, and hands are held.

This is not a waiting room.

This is life.

A mortal life is enough for meaning because meaning does not require infinity. It requires value.

The value is here: in the living, the suffering, the caring, the thinking, the courage, and the love.

Heaven is too distant.

Human life is here.

What Remains After God?

After God, what remains?

Everything real.

The sun still rises. The child still needs breakfast. The patient still needs treatment. The grieving still need comfort. The student still needs truth. The oppressed still need freedom. The lonely still need friendship. The body still needs care. The mind still needs honesty. The world still needs justice.

The absence of God does not remove these things.

It reveals them more clearly.

When God is removed, we can no longer hide behind divine mystery. We cannot say suffering is part of a plan. We cannot say hierarchy is sacred. We cannot say cruelty is obedience. We cannot say injustice will be corrected in another world.

We must look directly at life.

That direct gaze is the beginning of moral maturity.

A child is hungry. Feed the child. A woman is oppressed. Defend her freedom. A caste system humiliates. Reject it. A religion frightens children. Question it. A patient is suffering. Treat the pain. A belief is false. Do not teach it as truth. A life is brief. Love now.

This is what remains after God.

Not emptiness.

Reality.

The Human Being Is Enough

Religion often says human beings are not enough: not wise enough, not moral enough, not strong enough, not meaningful enough, not worthy enough without God.

But this insult has gone on long enough.

Human beings are imperfect. That is true. We are often selfish, fearful, tribal, violent, and foolish. But we are also capable of extraordinary compassion, courage, intelligence, tenderness, and moral growth.

We discovered germs and built medicine. We studied stars and learned our place in space. We challenged kings, priests, caste, slavery, and patriarchy. We wrote poetry and constitutions. We built hospitals and libraries. We comforted the dying. We cared for children not our own. We asked questions that frightened authority. We learned to say, "This ancient belief is wrong."

We are not perfect.

But we are not nothing.

The human being is enough to begin: enough to care, enough to think, enough to doubt, enough to learn, enough to help, enough to build morality, enough to create meaning, enough to live without God.

This does not mean human beings are gods.

It means gods were human creations, and the responsibility that we gave to them must now return to us.

Chapter Summary

This chapter brought together the central argument of the book.

When the soul disappears, the human being does not disappear. The self remains real as a living process created by brain, body, memory, emotion, language, relationship, and time.

When God disappears, morality does not disappear. Moral life remains grounded in the reality that conscious beings can suffer, flourish, be harmed, be protected, be humiliated, be loved, and be freed.

When heaven disappears, meaning does not disappear. Meaning remains in love, truth, work, beauty, justice, memory, service, and the difference one life makes in other lives.

The human being after God is not empty. The human being is natural, mortal, fragile, embodied, responsible, and worthy.

Science-based atheism does not reduce humanity. It removes false explanations and returns responsibility to human beings.

No soul is needed for mind.

No God is needed for morality.

No heaven is needed for meaning.

Key Takeaways

The absence of the soul does not erase the self. It makes the self a natural, living process.

The absence of God does not erase morality. It places morality on human suffering, dignity, compassion, justice, and reason.

The absence of heaven does not erase meaning. It makes this life more urgent and precious.

Human beings are natural, but not empty.

Human beings are bodies, but not mere objects.

Human beings are mortal, but not meaningless.

Human beings are responsible because no God will rescue us.

Science-based atheism is not moral despair. It is moral adulthood.

Transition to the Conclusion

We have reached the final step.

The soul was humanity's ancient explanation for the mind. God was humanity's ancient explanation for morality. Heaven was humanity's ancient answer to death.

But we no longer need these explanations.

The brain creates the self. Humanity creates ethics. This life creates meaning through love, truth, courage, and compassion.

The conclusion will now gather the whole argument into one final answer.

No soul needed.

No God required.

Sources

Antonio Damasio, *Self Comes to Mind: Constructing the Conscious Brain.*

Anil Seth, *Being You: A New Science of Consciousness.*

Michael Tomasello, *A Natural History of Human Morality.*

Frans de Waal, *The Bonobo and the Atheist.*

A.C. Grayling, *The God Argument.*

Paul Kurtz, *Forbidden Fruit: The Ethics of Humanism.*

Carl Sagan, *The Demon-Haunted World.*

Thaddeus Metz, "The Meaning of Life," *Stanford Encyclopedia of Philosophy.*

Matt Lutz, "Moral Naturalism," *Stanford Encyclopedia of Philosophy.*

M. Woźniak, "'I' and 'Me': The Self in the Context of Consciousness," 2018.

Conclusion: No Soul Needed, No God Required

Religion once told humanity that the soul made us human, that God made us moral, and that heaven made life meaningful.

For many centuries, this seemed believable. Human beings did not understand the brain. They did not understand consciousness. They did not understand how deeply memory, personality, awareness, emotion, and judgment depend on the living nervous system. They did not understand the natural roots of care, cooperation, fairness, empathy, and moral concern.

So they explained the mind with the soul.

They explained morality with God.

They explained death with heaven.

These explanations were emotionally powerful. They gave comfort to the grieving. They gave authority to moral rules. They gave people the hope that death was not final. They made the human being seem protected by something invisible and eternal.

But comfort is not the same as truth.

A belief may soothe fear and still fail as an explanation.

The central question of this book has been simple: if the soul is real, why does the brain control the mind?

A stroke can steal speech. Dementia can erase memory. A tumor can change judgment. Alcohol can alter restraint. Anesthesia can switch off consciousness. Brain injury can change personality. Sleep can make the self disappear every night.

These facts do not fit well with the idea of an independent soul. They fit much better with the idea that the mind is what the living brain does.

The soul was humanity's ancient answer to the mystery of the self. It grew from breath, dreams, death, memory, grief, and hope. It was understandable. It was human. But it was born before neuroscience.

Today, the soul no longer explains the evidence.

The brain does.

The Mind Does Not Need a Soul

To say that the brain creates the mind is not to insult the mind. It is to understand its fragile beauty.

The brain turns light into color, vibration into music, touch into tenderness, injury into pain, memory into identity, and another person's face into love, trust, or grief. This is not a small thing. It is astonishing.

The soul was meant to protect human wonder, but it is too simple for the richness of real human life. It cannot explain dementia. It cannot explain brain injury. It cannot explain anesthesia. It cannot explain why memory, language, personality, impulse, and awareness change when the brain changes.

A brain-based view explains more.

It also makes us more compassionate.

The depressed person is not simply weak. The addicted person is not simply sinful. The person with dementia is not spiritually empty. The patient with brain injury is not morally defective. The confused mind is not possessed by demons.

Human beings are biological, social, vulnerable, and shaped by causes. Understanding this does not remove dignity. It deepens care.

If the mind depends on the brain, then the brain matters. Sleep matters. Childhood matters. Mental health matters.

Dementia care matters. Addiction treatment matters. Stroke prevention matters. Pain relief matters. Compassion matters.

The soul asks us to imagine another world.

The brain asks us to care for this one.

Morality Does Not Need God

Religion also told humanity that without God, morality collapses.

But that fear is false.

Morality does not begin with divine command. It begins with conscious beings who can suffer.

A child's pain matters before any scripture speaks. A woman's dignity matters before any priest approves. A lower caste person's humanity matters even if sacred tradition denies it. A nonbeliever's life matters even if religion condemns disbelief.

A frightened child, a hungry family, a wounded patient, a grieving widow, an oppressed worker, and a silenced dissenter do not need heavenly permission before their suffering becomes morally important.

Their suffering matters because it is real.

Religion may encourage charity, kindness, service, and sacrifice. Many religious people are deeply moral, and that should be acknowledged honestly. But religion does not own goodness.

The moral value of feeding a hungry child is in the relief of hunger, not in the religious label attached to the food. The moral value of comforting the grieving is in the comfort, not in the doctrine. The moral value of protecting the weak is in the protection, not in the scripture.

Atheists can love, care, sacrifice, tell the truth, defend justice, and live responsibly without belief in God.

Goodness does not require heaven or hell. Fear of punishment is not the highest morality. Hope of reward is not the deepest compassion. Divine surveillance is not moral maturity.

A better morality asks: who is harmed, who is helped, whose dignity is protected, whose freedom is denied, what reduces suffering, what supports justice, and what allows human beings to flourish?

These questions do not require God.

They require honesty, courage, reason, and compassion.

Holy Commands Are Not Above Conscience

Divine-command morality fails because obedience can become dangerous when it replaces conscience.

A command does not become moral because it is called sacred. A cruelty does not become kindness because it is written in scripture. A hierarchy does not become just because it is ancient. A fear does not become love because it is preached.

History is full of examples. Slavery was defended by religion. Caste was protected as sacred order. Women were controlled in the name of divine law. Doubters, heretics, apostates, and blasphemers were punished to defend God.

This does not prove that every religious person is cruel. It proves something more precise and more important: when morality is placed under sacred authority, cruelty can be protected from criticism.

The best religious people are often better than the harshest parts of their religions. They soften cruel passages. They reinterpret old commands. They choose compassion over literal obedience. They allow conscience to correct scripture.

In doing so, they show that morality stands above command.

A better morality begins with real life. Begin with suffering. Protect dignity. Defend freedom. Seek justice. Follow evidence. Correct mistakes.

That morality is not perfect because human beings are not perfect. But it can improve, and that is its strength.

A morality that can admit error is better than a morality that calls ancient error eternal.

Meaning Does Not Need Heaven

Religion's final shelter is heaven.

It says that without immortality, life becomes meaningless.

But a flower is not worthless because it fades. A song is not meaningless because it ends. A childhood is not empty because it passes. A friendship is not fake because death may part it. A human life does not lose its value because it is brief.

Meaning is not measured by endless duration.

Meaning is measured by value.

Love matters because it is lived. Kindness matters because it reduces suffering. Truth matters because falsehood imprisons the mind. Justice matters because victims cannot wait for heaven. Beauty matters because conscious beings can experience it. Memory matters because one life can continue as influence inside other lives.

Heaven may comfort grief, but grief does not prove heaven. The longing to see the dead again is deeply human, but longing is not evidence.

A mortal life can still be deep, beautiful, serious, loving, and worth living.

In fact, mortality can make life more precious.

Because time is limited, kindness should not be postponed. Because people die, love should be spoken. Because there may be no afterlife, this life should not be wasted. Because no heaven has been shown to exist, justice must be pursued here.

This life is not a waiting room. This is where children need food, patients need care, women need freedom, caste must be rejected, truth must be defended, love must be expressed, suffering must be reduced, and justice must be done.

This is where meaning is made.

What Remains

After the soul, what remains?

The living brain, creating a self.

After God, what remains?

Human beings, responsible for morality.

After heaven, what remains?

This brief life, filled with value because it can be loved, harmed, healed, wasted, defended, and shared.

The absence of God does not make the world empty.

It makes the world ours.

No divine hand may stop the bullet, cure the disease, feed the hungry, protect the child, end caste, free women, defend dissenters, or comfort the dying.

But human hands can do some of this. Human minds can. Human courage can. Human science can. Human compassion can. Human law can. Human solidarity can.

That is not despair.

That is responsibility.

And responsibility is one of the deepest forms of human dignity.

The Final Answer

This book has not argued that human beings are perfect.

We are not.

We are often selfish, fearful, tribal, cruel, irrational, and easily misled. But we are also capable of tenderness, reason, courage, science, art, medicine, forgiveness, justice, and moral growth.

We can learn. We can change. We can question inherited beliefs. We can reject sacred cruelty. We can leave false comfort. We can build meaning without illusion. We can protect dignity without divine command. We can love without believing love is eternal. We can live honestly without pretending death is a doorway.

That is not a small achievement.

It is one of the noblest achievements of the human mind.

Religion says the soul makes us human.

But the brain creates the self.

Religion says God makes us moral.

But humanity creates ethics.

Religion says heaven makes life meaningful.

But this one real life is meaningful enough.

No soul is needed for love.

No God is needed for kindness.

No heaven is needed for meaning.

We need truth.

We need courage.

We need compassion.

We need justice.

We need freedom.

We need one another.

That is not a fall from human dignity.

That is human dignity standing on its own feet.

Mind without soul.

Morality without God.

Meaning without immortality.

No God required.

Sources

Antonio Damasio, *Self Comes to Mind: Constructing the Conscious Brain.*

Anil Seth, *Being You: A New Science of Consciousness.*

Stanislas Dehaene, *Consciousness and the Brain.*

Michael Tomasello, *A Natural History of Human Morality.*

Frans de Waal, *The Bonobo and the Atheist.*

A.C. Grayling, *The God Argument.*

Carl Sagan, *The Demon-Haunted World.*

Paul Kurtz, *Forbidden Fruit: The Ethics of Humanism.*

Thaddeus Metz, "The Meaning of Life," *Stanford Encyclopedia of Philosophy.*

Matt Lutz, "Moral Naturalism," *Stanford Encyclopedia of Philosophy.*

Appendix A: Common Religious Claims and Clear Answers

Many people do not leave religious belief because of one argument.

They leave slowly.

A question appears. Then another. A doubt grows. A comforting belief becomes less convincing. A fear remains. A childhood teaching still echoes. A religious claim that once felt certain begins to feel weak.

This appendix is written for that process.

It answers common religious claims about the soul, morality, God, heaven, death, and meaning in clear language. The goal is not to win an argument by insult. The goal is to answer honestly, calmly, and firmly.

Religion often survives because it gives short, emotionally powerful answers.

“You have a soul.”

“God gives morality.”

"Heaven gives life meaning."

"Without God, anything is permitted."

"Atheists have no basis for right and wrong."

These claims sound strong until they are examined.

This book has examined them in detail. This appendix gathers the answers in a simpler form.

Claim 1: If There Is No Soul, What Makes Me "Me"?

You are not made into "you" by an invisible soul. You are made by your living brain, body, memory, emotions, relationships, language, habits, history, and choices.

The self is not a fixed object hidden inside the body. It is a living process. You feel like the same person over time because your brain connects memory, body signals, personal history, social identity, and future plans into one continuing story. That story can change, grow, break, heal, and revise itself. But it is still real.

A life story can be real without being immortal. A person can change and still have continuity. A self can be natural and still be precious.

The absence of a soul does not mean you are nothing. It means you are a living, changing, embodied human being.

Claim 2: If the Brain Creates Love, Is Love Fake?

No.

Love is not fake because the brain participates in it.

A rose is not fake because it grows from soil, sunlight, water, and genes. A rainbow is not fake because it comes from light and water droplets. Music is not fake because it depends on vibrating air and a hearing brain.

Love has biology inside it, but it is not "only" biology in any empty sense. Love includes attachment, memory, trust, touch, time, sacrifice, concern, tenderness, desire, loyalty, and shared life.

Understanding the brain does not destroy love. It helps us understand how precious love is.

The brain allows us to recognize faces, remember voices, feel attachment, grieve loss, and care about another person's well-being. That does not make love smaller.

It makes the brain more astonishing.

Claim 3: If Consciousness Is Not Fully Explained, Does That Leave Room for God?

Not in the way religion claims.

Consciousness remains one of the hardest problems in science and philosophy. We do not yet have a complete explanation of how subjective experience arises from brain activity.

But "not fully explained" does not mean "therefore God."

Many things were once not fully explained: lightning, disease, epilepsy, dreams, birth, death, and evolution. Religion often placed God or spirits inside those gaps. As knowledge grew, the supernatural explanations retreated.

The honest answer to an unsolved question is, "We do not yet fully know."

The dishonest answer is, "We do not know, therefore my religion is true."

Consciousness is mysterious, but it clearly depends deeply on the brain. It changes with sleep, anesthesia, drugs, oxygen failure, seizures, dementia, coma, and brain injury.

That is exactly what we would expect if consciousness is brain-based.

It is not good evidence for an immortal soul.

Claim 4: If the Soul Is Not Real, What Happens After Death?

The honest answer is that we have no good evidence that personal consciousness continues after the brain permanently dies.

Everything we know about ordinary consciousness shows deep dependence on the living brain. When the brain is injured, consciousness can change. When the brain is anesthetized, consciousness can disappear for a time. When the brain degenerates, memory and personality can change. When the brain permanently stops functioning, there is no reliable evidence that the self continues elsewhere.

This is painful.

But pain does not make a belief true.

The absence of an afterlife does not erase love, memory, legacy, grief, gratitude, or the effect one life has on other lives.

A person may not live forever as a soul, but a life can still matter through influence, children, ideas, kindness, work, memory, and the changes it leaves in others.

That is not immortality.

But it is real.

Claim 5: If There Is No God, Why Be Good?

Because other beings can suffer.

Because cruelty wounds.

Because trust matters.

Because children need protection.

Because honesty supports human life.

Because kindness reduces suffering.

Because justice protects the vulnerable.

Because freedom and dignity matter to conscious beings.

God is not needed for any of this.

If the only reason a person avoids cruelty is fear of God, that is not deep morality. That is fear of punishment.

A mature morality asks not, "What will happen to me if I am caught?"

It asks, "What will happen to others if I do this?"

Morality begins when we recognize that other beings have inner lives, just as we do. They can feel pain, fear, humiliation, grief, hunger, loneliness, and hope.

That is enough reason to care.

Claim 6: Without God, Is Anything Permitted?

No.

Without God, everything is not permitted. Without God, actions must still be judged by their effect on conscious beings.

Murder is wrong because it destroys a life. Rape is wrong because it violates autonomy, dignity, safety, and the body. Slavery is wrong because it crushes freedom and personhood. Caste humiliation is wrong because birth does not determine human worth. Patriarchy is wrong because women are full human beings. Religious persecution is wrong because conscience must be free. Child abuse is wrong because children are vulnerable and can be deeply harmed.

None of these moral truths require God.

They require attention to suffering, dignity, freedom, and harm.

A godless morality is not "anything goes."

It is "nothing is above moral examination."

That includes scripture, priesthood, tradition, and God-claims.

Claim 7: Are Atheists Less Moral Than Religious People?

No.

Atheists can be kind, honest, loving, generous, faithful, brave, and responsible. Religious people can also be kind, honest, loving, generous, faithful, brave, and responsible.

Belief in God does not automatically make a person good. Disbelief in God does not automatically make a person bad.

People are shaped by empathy, upbringing, culture, law, education, personality, social pressure, fear, reason, habit, and moral reflection. Religion can sometimes encourage moral behavior. It can also encourage prejudice, obedience, fear, and cruelty.

Atheism alone is not a complete moral system. It simply rejects belief in God.

But atheism combined with humanism, reason, compassion, and respect for dignity can support a strong moral life.

The real question is not whether a person believes in God. The real question is whether the person reduces suffering, protects dignity, respects freedom, values truth, and treats others with fairness.

Claim 8: If Morality Evolved, Does That Make It Selfish or Meaningless?

No.

The fact that morality has evolutionary roots does not make morality selfish or meaningless.

Human beings evolved as social animals. We survived through care, cooperation, bonding, trust, warning, sharing, and group life. These roots helped make morality possible.

But morality did not stop there. Human reason can expand moral concern beyond family, tribe, religion, caste, nation, and species. We can understand that strangers suffer as we do. We can recognize that old customs may be unjust. We can use evidence to reduce harm. We can correct inherited prejudice.

The origin of something does not determine its value.

Music may have biological roots, but music is still beautiful. Language evolved, but poetry is still meaningful. Love has biological roots, but love is still real.

Morality may have natural roots, but justice, compassion, and dignity still matter.

Claim 9: If Morality Is Human-Made, Is It Just Opinion?

No.

Human-made does not mean worthless or arbitrary.

Medicine is human-made, but it can heal. Law is human-made, but it can protect rights. Language is human-made, but it can carry truth. Democracy is human-made, but it can defend freedom.

Morality can be human-made and still be serious because it deals with real things: pain, fear, hunger, oppression, dignity, freedom, trust, and well-being.

Of course, moral reasoning can be difficult. People can disagree. Cultures can be wrong. Societies can defend cruelty. That is why morality must remain open to criticism, evidence, and improvement.

Divine-command morality often claims certainty, but certainty can protect error.

Human morality admits responsibility. It says: we must examine consequences, listen to victims, correct mistakes, and improve.

That is not weakness.

That is moral maturity.

Claim 10: Where Does Human Dignity Come From Without God?

Human dignity comes from the reality of conscious life.

A person has inner experience. A person can suffer. A person can hope. A person can fear. A person can be humiliated. A person can be loved. A person can flourish or be crushed.

That is enough to make the person morally important.

Dignity does not need to be granted by religion. If dignity depends on God, then people outside that religion may be treated as less worthy. History shows how dangerous that can be.

A stronger foundation says every person matters because every person is a conscious, vulnerable human being.

A newborn has dignity before belief. A woman has dignity before scripture. A lower caste person has dignity before tradition. An atheist has dignity before religious approval. A person with dementia has dignity even when memory fades. A prisoner has dignity even after wrongdoing.

Human dignity is not a reward for correct belief.

It is the moral status of a conscious human life.

Claim 11: If There Is No Heaven, What Is the Point of Life?

The point of life does not have to come from heaven.

It can come from love, truth, work, service, friendship, beauty, courage, learning, healing, justice, creativity, and reducing suffering.

A life does not need to last forever to matter.

A song matters though it ends. A meal matters though it is eaten. A childhood matters though it passes. A friendship matters though death may part it. A patient comforted near death matters though the patient dies.

Meaning is not measured by endless duration.

Meaning is measured by value.

If this is the only life we know we have, then it becomes more precious, not less.

This life is not a waiting room. This is where love must be expressed, suffering must be reduced, and justice must be defended.

Claim 12: Does Atheism Make Life Cold and Empty?

No.

Atheism removes supernatural beliefs. It does not remove love, beauty, music, friendship, kindness, humor, science, courage, or compassion.

The morning still comes. Children still laugh. Food still tastes good. Books still open minds. The sick still need care. The grieving still need comfort. The oppressed still need freedom. The truth still matters.

A godless world is not empty.

It is real.

And reality is full of meaning when we stop looking above the clouds and begin looking carefully at life itself.

Science-based atheism can make life deeper because it removes false comfort and gives us honest responsibility.

No God will fix this world for us.

That means we must do the work ourselves.

That is not emptiness.

That is moral adulthood.

Claim 13: If There Is No Divine Justice, How Can We Accept Unfairness?

We may not be able to accept it easily.

That is exactly why we must fight it here.

Religion often says that injustice will be corrected after death. The wicked will be punished. The good will be rewarded. The suffering will be healed in heaven.

This belief can comfort people, but it can also weaken urgency. If heaven will fix injustice later, people may tolerate too much injustice now.

A secular view says there is no reliable evidence that justice will be completed after death.

That is painful.

But it makes justice in this life more urgent.

If a child is hungry, we must feed the child. If a woman is oppressed, we must defend her freedom. If caste humiliates people, we must reject caste. If religious violence harms communities, we must resist it. If a patient suffers, we must treat pain.

If there is no divine justice waiting, human justice matters more.

Claim 14: Is Religion Necessary to Keep Society Stable?

No.

Societies need trust, law, education, fairness, compassion, public health, rights, accountability, and shared responsibility.

Religion can sometimes support community and charity. But religion can also divide societies, protect hierarchy, frighten children, suppress women, punish dissent, and create hostility toward outsiders.

Social stability does not require belief in God. It requires decent institutions and moral habits.

A society is healthier when people are taught to think clearly, respect evidence, protect freedom, care for the vulnerable, and treat all citizens with equal dignity.

Fear-based religious obedience is not the same as social health.

A stable society built on fear may look orderly, but it is not free.

A better society is built on justice, dignity, education, and compassion.

Claim 15: If There Is No Soul, Are Humans Just Animals?

Humans are animals.

But the word “just” is misleading.

We are animals with language, memory, imagination, culture, moral reasoning, science, art, medicine, grief, humor, and self-reflection.

Being animals does not make us worthless. It connects us to life.

The fact that we evolved does not make love false. It does not make justice meaningless. It does not make suffering unimportant.

A human being is a natural creature, but nature is not empty.

The human brain is part of nature, and it can create poetry, surgery, mathematics, compassion, music, and moral courage.

We are not less wondrous because we are animals.

We are animals who can know that we are animals.

That is astonishing.

Claim 16: Does Science Explain Everything?

No.

Science does not explain everything yet. It may never explain every question in a final or complete way.

Consciousness remains difficult. Free will remains debated. Morality remains complex. Death remains emotionally painful. Meaning remains something each life must build.

But the fact that science does not explain everything does not mean religion is true.

A gap in knowledge is not evidence for God.

Science is honest about uncertainty. Religion often fills uncertainty with authority.

A mature mind can say, “I do not know,” without adding, “Therefore God.”

That humility is not weakness.

It is intellectual honesty.

Claim 17: What Should Replace Religion?

Nothing needs to replace religion as another dogma.

But several things can replace the functions religion often served.

Science can replace supernatural explanation. Humanism can replace divine-command morality. Philosophy can replace unquestioned doctrine. Community can replace religious belonging. Art and music can replace ritual beauty. Therapy and friendship can replace confession and spiritual fear. Medicine can replace prayer as treatment. Justice can replace divine judgment.

Meaningful work can replace heavenly purpose. Love can replace fear. Truth can replace faith.

The goal is not to build another church of atheism.

The goal is to build a life based on reality, compassion, freedom, and responsibility.

Claim 18: What Is the Final Answer?

The final answer is simple.

The soul was an ancient explanation for the mind.

The brain explains the mind better.

God was an ancient explanation for morality.

Human life explains morality better.

Heaven was an ancient answer to death.

This life is meaningful enough without heaven.

No soul is needed to love.

No God is needed to be kind.

No heaven is needed for life to matter.

We need truth.

We need courage.

We need compassion.

We need justice.

We need freedom.

We need one another.

That is enough.

No God required.

Sources

Antonio Damasio, *Self Comes to Mind: Constructing the Conscious Brain.*

Anil Seth, *Being You: A New Science of Consciousness.*

Michael Tomasello, *A Natural History of Human Morality.*

Frans de Waal, *The Bonobo and the Atheist.*

Greg M. Epstein, *Good Without God.*

A.C. Grayling, *The God Argument.*

Paul Kurtz, *Forbidden Fruit: The Ethics of Humanism.*

Carl Sagan, *The Demon-Haunted World.*

Bertrand Russell, "Why I Am Not a Christian."

Thaddeus Metz, "The Meaning of Life," *Stanford Encyclopedia of Philosophy*.

Stanford Encyclopedia of Philosophy, "Moral Naturalism."

Stanford Encyclopedia of Philosophy, "Death."

Internet Encyclopedia of Philosophy, "Divine Command Theory."

Appendix B: Suggested Reading and Sources

Readers do not need to study these sources to understand this book. They are included for those who want to go deeper.

This book was written for general readers, not specialists. Its purpose is to explain, in clear and human language, why the mind does not require a soul, why morality does not require God, and why meaning does not require heaven.

The books, articles, and reference sources listed below helped shape the scientific, philosophical, and ethical arguments in this book. Some are written for general readers. Others are more technical. Together, they provide a reliable path for readers who want to explore these questions further.

This list is not meant to be exhaustive. It is meant to show that the arguments in this book rest on serious work in neuroscience, psychology, philosophy, ethics, anthropology, and humanism.

Mind, Brain, and the Self

Antonio Damasio, *Descartes' Error: Emotion, Reason, and the Human Brain.*

Antonio Damasio, *Self Comes to Mind: Constructing the Conscious Brain.*

Anil Seth, *Being You: A New Science of Consciousness.*

Stanislas Dehaene, *Consciousness and the Brain.*

Michael S. Gazzaniga, *Who's in Charge? Free Will and the Science of the Brain.*

Oliver Sacks, *The Man Who Mistook His Wife for a Hat.*

Daniel L. Schacter, *The Seven Sins of Memory.*

Christof Koch, *The Feeling of Life Itself.*

David J. Chalmers, *The Conscious Mind.*

Patrick McNamara, *The Neuroscience of Religious Experience.*

M. Woźniak, "'I' and 'Me': The Self in the Context of Consciousness," 2018.

H. J. Markowitsch and A. Staniloiu, "Memory and Self: Neuroscientific Landscapes," *Frontiers in Psychology*, 2013.

Arnaud D'Argembeau, "Memory, Future Thinking, and the Self," *Memory & Cognition*, 2021.

A. Monti et al., "The Inside of Me: Interoceptive Constraints on the Concept of Self," *Neuroscience & Biobehavioral Reviews*, 2021.

These sources are useful for understanding how memory, body signals, emotion, identity, self-awareness, and consciousness are connected to the brain. They support one of the central arguments of this book: the self is real, but it is not an immortal soul. It is a living process created by the brain, body, memory, emotion, and relationships.

Brain Injury, Dementia, Anesthesia, and Disorders of Consciousness

Hanna Damasio et al., "The Return of Phineas Gage: Clues About the Brain from the Skull of a Famous Patient," *Science*, 1994.

P. H. M. de Freitas et al., "E.L., a Modern-Day Phineas Gage: Revisiting Frontal Lobe Injury," *Dementia & Neuropsychologia*, 2022.

S. M. Sheppard and S. Sebastian, "Diagnosing and Managing Post-Stroke Aphasia," *Expert Review of Neurotherapeutics*, 2020.

A. Terracciano et al., "Personality and Alzheimer's Disease: An Integrative Review," *Personality Disorders*, 2019.

Y. Nir and G. Tononi, "Dreaming and the Brain: From Phenomenology to Neurophysiology," *Trends in Cognitive Sciences*, 2010.

B. Baird et al., "The Cognitive Neuroscience of Lucid Dreaming," *Neuroscience & Biobehavioral Reviews*, 2019.

S. Scarpelli et al., "The Functional Role of Dreaming in Emotional Processes," *Frontiers in Psychology*, 2019.

J. Montupil et al., "The Nature of Consciousness in Anaesthesia," *British Journal of Anaesthesia*, 2023.

A. G. Hudetz, "Disconnecting Consciousness: Is There a Common Anesthetic End Point?" *Anesthesia & Analgesia*, 2016.

J. T. Giacino et al., "Comprehensive Systematic Review Update Summary: Disorders of Consciousness," *Neurology*, 2018.

B. L. Edlow et al., "Recovery from Disorders of Consciousness," *Nature Reviews Neurology*, 2020.

Centers for Disease Control and Prevention, "About Sleep," 2024.

These sources support the clinical argument that the mind depends deeply on the brain. Brain injury can affect personality and judgment. Stroke can damage speech and understanding. Dementia can affect memory, identity, and behavior. Anesthesia can profoundly alter consciousness. Disorders of consciousness show how closely awareness depends on brain function.

These facts are among the strongest reasons to question the idea of a separate, independent soul.

Consciousness and Scientific Theories of Mind

Cogitate Consortium et al., "Adversarial Testing of Global Neuronal Workspace and Integrated Information Theories of Consciousness," *Nature*, 2025.

G. Friedman et al., "The Current of Consciousness: Neural Correlates and Clinical Aspects of Consciousness," 2023.

Stanford Encyclopedia of Philosophy, "Consciousness."

Stanford Encyclopedia of Philosophy, "Physicalism."

Stanford Encyclopedia of Philosophy, "Naturalism."

Consciousness remains one of the hardest problems in science and philosophy. No final theory has won. But modern consciousness research is active, empirical, and centered on brain function.

The important point for this book is simple: mystery is not evidence for God. An unanswered scientific question is not proof of the soul.

Free Will, Choice, and Responsibility

Daniel Dennett, *Freedom Evolves.*

Michael S. Gazzaniga, *Who's in Charge? Free Will and the Science of the Brain.*

Robert Sapolsky, *Behave: The Biology of Humans at Our Best and Worst.*

Robert Sapolsky, *Determined: A Science of Life Without Free Will.*

Sam Harris, *Free Will.*

M. Brass, P. Haggard, and others, "Why Neuroscience Does Not Disprove Free Will," *Neuroscience & Biobehavioral Reviews*, 2019.

C. D. Frith and P. Haggard, "Volition and the Brain: Revisiting a Classic Experimental Study," *Trends in Neurosciences*, 2018.

A. Schurger et al., "What Is the Readiness Potential?" *Trends in Cognitive Sciences*, 2021.

J. Shepherd, "Scientific Challenges to Free Will and Moral Responsibility," *Philosophy Compass*, 2015.

A. Bigenwald and F. Chambon, "Criminal Responsibility and Neuroscience: No Revolution Yet," *Frontiers in Psychology*, 2019.

Stanford Encyclopedia of Philosophy, "Free Will."

These sources help explain why human choice should not be understood as a ghost-like soul choosing outside nature. Choices arise from brain function, memory, emotion, habit, development, health, culture, and circumstance.

At the same time, responsible scholarship is careful. Neuroscience has not simply "disproved free will" in one easy stroke. A better conclusion is that human freedom is natural, limited, embodied, and connected to capacities such as reflection, impulse control, learning, and response to reasons.

Morality, Cooperation, and Evolution

Michael Tomasello, *A Natural History of Human Morality*.

Michael Tomasello, "Origins of Human Cooperation and Morality," *Annual Review of Psychology*, 2013.

Frans de Waal, *The Bonobo and the Atheist*.

Frans de Waal, *Primates and Philosophers*.

Sarah F. Brosnan and Frans B. M. de Waal, "Evolution of Responses to (Un)Fairness," *Science*, 2014.

P. Limone and G. Toto, "Origin and Development of Moral Sense: A Systematic Review," 2022.

K. A. Dunfield, "A Construct Divided: Prosocial Behavior as Helping, Sharing, and Comforting," *Frontiers in Psychology*, 2014.

Oliver Scott Curry, Daniel Austin Mullins, and Harvey Whitehouse, "Is It Good to Cooperate? Testing the Theory of Morality-as-Cooperation in 60 Societies," *Current Anthropology*, 2019.

Luca Surian, "Core Moral Concepts and the Sense of Fairness in Human Infants," 2025.

Richard Joyce, *The Evolution of Morality.*

These sources support the argument that morality has natural and social roots. Care, cooperation, fairness, helping, sharing, empathy, guilt, gratitude, and compassion did not begin with scripture. They grew from social life.

Religion may shape morality. It may encourage morality. It may also distort morality. But it did not create the basic human need for care, fairness, trust, cooperation, and protection from harm.

Atheism, Humanism, and Morality Without God

Greg M. Epstein, *Good Without God.*

Paul Kurtz, *Forbidden Fruit: The Ethics of Humanism.*

A.C. Grayling, *The God Argument.*

Phil Zuckerman, *Society Without God.*

Richard Dawkins, *The God Delusion.*

Bertrand Russell, "Why I Am Not a Christian."

Carl Sagan, *The Demon-Haunted World.*

Robert McKay and Harvey Whitehouse, "Religion and Morality," *Psychological Bulletin*, 2015.

T. Ståhl, "The Amoral Atheist? A Cross-National Examination of Cultural, Motivational, and Cognitive Antecedents of Disbelief, and Their Implications for Morality," *PLOS ONE*, 2021.

W. M. Gervais et al., "Global Evidence of Extreme Intuitive Moral Prejudice Against Atheists," *Nature Human Behaviour*, 2017.

These sources are useful for understanding secular ethics, humanism, and the prejudice often directed against atheists. The evidence does not support the crude claim that atheists cannot be moral. Atheists can be kind, honest, compassionate, responsible, and deeply concerned with justice.

Atheism alone is not a complete moral system. But atheism joined with humanism, reason, compassion, and respect for dignity can support a strong moral life.

Divine Command Theory and the Problem of Holy Morality

Plato, *Euthyphro*.

Internet Encyclopedia of Philosophy, "Divine Command Theory."

Stanford Encyclopedia of Philosophy, "Theological Voluntarism."

Stanford Encyclopedia of Philosophy, “Religion and Morality.”

H. Van Eyghen, “Debunking Divine Command Theory,” *Religions*, 2023.

Peter Singer, *Practical Ethics.*

A.C. Grayling, *The God Argument.*

Bertrand Russell, “Why I Am Not a Christian.”

These sources support the critique of divine-command morality. The central question is ancient and powerful: is something good because God commands it, or does God command it because it is good?

If something is good only because God commands it, morality becomes obedience to power. If God commands something because it is already good, then goodness does not depend on God.

This is one of the strongest philosophical reasons to reject the claim that morality requires God.

Human Dignity, Human Rights, Justice, and Well-Being

Peter Singer, *Practical Ethics.*

Martha Nussbaum, *Creating Capabilities.*

Amartya Sen, *The Idea of Justice.*

John Rawls, *A Theory of Justice.*

United Nations, *Universal Declaration of Human Rights.*

Stanford Encyclopedia of Philosophy, "Dignity."

Stanford Encyclopedia of Philosophy, "Human Rights."

Stanford Encyclopedia of Philosophy, "Well-Being."

Internet Encyclopedia of Philosophy, "Sen's Capability Approach."

These sources support the positive moral framework used in this book. A better morality can be built on human dignity, harm reduction, freedom, justice, equality, evidence, and well-being.

Human dignity does not depend on religion. A person has worth before belief, caste, gender, wealth, status, usefulness, or social approval. A secular morality asks what reduces suffering, protects freedom, respects dignity, and helps human beings flourish.

Meaning, Death, and Life Without Heaven

Irvin D. Yalom, *Staring at the Sun: Overcoming the Terror of Death.*

Viktor E. Frankl, *Man's Search for Meaning.*

Albert Camus, *The Myth of Sisyphus.*

Ernest Becker, *The Denial of Death.*

Bertrand Russell, "A Free Man's Worship."

Carl Sagan, *The Demon-Haunted World.*

Thaddeus Metz, "The Meaning of Life," *Stanford Encyclopedia of Philosophy.*

Steven Luper, "Death," *Stanford Encyclopedia of Philosophy.*

I. J. Peltomäki, "Philosophy of Meaning in Life in the Light of Finitude," 2023.

M. Svet et al., "Applying Terror Management Theory to Patients With Life-Threatening Illnesses," 2023.

These sources help support the argument that life can be meaningful without immortality. A finite life can still contain love, courage, work, beauty, truth, service, justice, and care.

Meaning does not require heaven. It requires value.

A flower is not worthless because it fades. A song is not meaningless because it ends. A human life is not empty because it is brief.

Religion, Death, Dreams, and the Origin of the Soul

Edward B. Tylor, *Primitive Culture.*

Patrick McNamara, *The Neuroscience of Religious Experience*.

E. J. Wamsley, "How the Brain Constructs Dreams," *eLife*, 2020.

National Center for Biotechnology Information, *The Possible Functions of REM Sleep and Dreaming*.

D. M. Varisco, "The End of Life, The Ends of Life: An Anthropological View," *Journal of the Islamic Medical Association of North America*, 2012.

R. E. Watson-Jones et al., "Does the Body Survive Death? Cultural Variation in Beliefs About Life Everlasting," *Cognitive Science*, 2016.

These sources are useful for understanding why belief in the soul became so powerful. Breath, dreams, death, grief, memory, and fear all helped make the soul seem real to ancient people.

Soul-belief was understandable as an early human explanation.

But an ancient explanation is not modern evidence.

Helpful Reference Sources

Stanford Encyclopedia of Philosophy.

Internet Encyclopedia of Philosophy.

National Center for Biotechnology Information.

National Institutes of Health.

Centers for Disease Control and Prevention.

United Nations human rights documents.

Peer-reviewed journals in neuroscience, psychology, philosophy, anthropology, and moral science.

These reference sources are useful because they provide careful, review-based information. Readers who want to study these topics further should begin with high-quality sources rather than social media posts, apologetic websites, or polemical videos.

Final Note to the Reader

The purpose of this appendix is not to overwhelm the reader with names and titles. The purpose is to show that the argument of this book does not rest on anger, mockery, or personal opinion. It rests on a wide body of scientific, philosophical, historical, and ethical work.

The central message remains simple.

The brain explains the mind better than the soul.

Human life explains morality better than God.

This world explains meaning better than heaven.

No soul is needed.

No God is required.

Thank You and Please Review the Book

Thank you for taking the time to read this book. I sincerely hope it has encouraged you to think more deeply, question more honestly, and look at religion, science, and human belief with greater clarity.

My goal in writing the **No God Required Series** is simple: to present serious, science-based, and easy-to-understand ideas about God, religion, nature, morality, and human meaning. These books are written for readers who value evidence, reason, intellectual honesty, and the courage to ask difficult questions.

If you found this book thoughtful, helpful, or worth reading, I would be truly grateful if you could leave a brief and honest review. Even a few sincere lines can make a meaningful difference. Your review helps other readers discover this series and encourages more open, rational, and evidence-based discussion about some of life's deepest questions.

Thank you again for your time, your trust, and your support.

Warm regards,
Dr. Prabhat Das

OTHER BOOKS IN THIS SERIES

No God Required: How Science Explains Reality Without the Divine is a bold, clear, science-based series by **Dr. Prabhat Das**. It reveals why humans created gods, how science made supernatural explanations unnecessary, and how meaning, morality, courage, and wonder can flourish without the divine.

<u>The 7 Books in This Series: -</u>

Book 1: **Why and How Humans Created God**

Book 2: **How Science Made God Unnecessary**

Book 3: **The Universe Without a Creator**

Book 4: **Life Without a Designer**

Book 5: **Mind Without Soul, Morality Without God**

Book 6: **The Failure of the Personal God**

Book 7: **The Scientific Case for Atheism**

www.ingramcontent.com/pod-product-compliance
Ingram Content Group UK Ltd.
Pitfield, Milton Keynes, MK11 3LW, UK
UKHW020839290926
15470UKWH00004B/1431

9 798951 067128